List of Sponsors (revised on February 2, 2018)

EAST JAPAN RAILWAY COMPANY
KAJIMA CORPORATION
SHIMIZU CORPORATION
TAISEI CORPORATION
TOKYO SOIL RESEARCH CO., LTD.
FUDO TETRA CORPORATION
RAITO KOGYO CO., LTD.
CHIBA ENGINEERING CORP.
CHUO KAIHATSU CORPORATION
DIA CONSULTANTS CO., LTD.
GEO-LABO CHUBU
GIKEN LTD.
JAPAN PORT CONSULTANTS, LTD.
KINJO RUBBER CO., LTD.*
KISO-JIBAN CONSULTANTS CO., LTD.
NAKANIHON ENGINEERING CONSULTANTS CO., LTD.
NEWJEC INC.
NIPPON KOEI CO., LTD.
NITTOC CONSTRUCTION CO., LTD.
OBAYASHI CORPORATION
OKUMURA CORPORATION
OYO CORPORATION
SUMITOMO MITSUI CONSTRUCTION CO., LTD.
TOKYO ELECTRIC POWER SERVICES CO., LTD. (TEPSCO)*
TOKYU CONSTRUCTION
TOYO CONSTRUCTION CO., LTD.
YBM CO., LTD.
ASANO TAISEI KISO ENGINEERING CO., LTD.
CHUO FUKKEN CONSULTANTS CO., LTD.
CHEMICAL GROUTING CO., LTD.
CTI ENGINEERING CO., LTD.
INTEGRATED GEOTECHNOLOGY INSTITUTE LIMITED
KAWASAKI GEOLOGICAL ENGINEERING CO., LTD.
KUMAGAI GUMI CO., LTD.
MAEDA CORPORATION
NISHIMATSU CONSTRUCTION CO., LTD.
ONODA CHEMICO CO., LTD.
SANSHIN CORPORATION
TEKKEN CORPORATION
TENOX CORPORATION
KANSAI GEO AND ENVIRONMENT RESEARCH CENTER
HOKKAIDO SOIL RESEARCH CO-OPERATION
CHIYODA ENGINEERING CONSULTANTS CO., LTD.*
CHODAI CO., LTD.*
CHUDEN ENGINEERING CONSULTANTS
COASTAL DEVELOPMENT INSTITUTE OF TECHNOLOGY
EIGHT-JAPAN ENGINEERING CONSULTANTS INC.
ETERNAL PRESERVE*
FUJITA CORPORATION
FUKKEN CO., LTD.
HAZAMA ANDO CORPORATION
HOJUN CO., LTD.
HOKUKOKU CHISUI CO., LTD.
JAPAN CONSERVATION ENGINEERS & CO., LTD.
MARUI & CO., LTD.
NIKKEN SEKKEI CIVIL ENGINEERING LTD.
NIPPON ENGINEERING CONSULTANTS CO., LTD.*
NiX CO., LTD.
ORIENTAL CONSULTANTS CO., LTD.*
ORIENTAL CONSULTANTS GLOBAL CO., LTD.*
THE OVERSEAS COASTAL AREA DEVELOPMENT INSTITUTE OF JAPAN (OCDI)
PACIFIC CONSULTANTS CO., LTD.
PENTA-OCEAN CONSTRUCTION CO., LTD.
SATO KOGYO CO., LTD.
SERVICE CENTER OF PORT ENGINEERING (SCOPE)
SOIL AND ROCK ENGINEERING CO., LTD.
TAKENAKA CIVIL ENGINEERING & CONSTRUCTION CO., LTD.
TAKENAKA CORPORATION
TOA CORPORATION
TOA GROUT CO., LTD.*
WESCO CO., LTD.
YONDEN CONSULTANTS CO., LTD.*

* No AD.

JAPANESE GEOTECHNICAL SOCIETY STANDARDS

Geotechnical and Geoenvironmental Investigation Methods

Vol. 3

THE JAPANESE GEOTECHNICAL SOCIETY

Foreword to Vol. 3

The project to translate and publish English language versions of "The Japanese Geotechnical Society Standards" was commenced in 2014 as a special project of former JGS President Ikuo Towhata, with the aim of publishing the English standards in 3 volumes, Vol. 1 to Vol. 3, over 3 years (2015 to 2017). Vol. 1 (published in 2015, containing 20 "Geotechnical and Geoenvironmental Investigation Methods" and 20 "Laboratory Testing Standards of Geomaterials") and Vol. 2 (published in 2016, containing 20 "Geotechnical and Geoenvironmental Investigation Methods" and 20 "Laboratory Testing Standards of Geomaterials") have been very well received, in particular by overseas government organizations, and are being used both in Japan and overseas. By bringing these excellent standards of the Japanese Geotechnical Society to the world, these volumes are contributing to the spread of technology in "Geotechnical and Geoenvironmental Investigation Methods" and "Laboratory Testing of Geomaterials". Vol. 3 contains 25 new "Geotechnical and Geoenvironmental Investigation Methods" and 30 new "Laboratory Testing Standards of Geomaterials". It is our wish that this volume will be widely used together with Vol. 1 and Vol. 2 in many fields in engineering practice, in education, and in research.

Very many donations have been received for the project to translate and publish English language versions of the Japanese Geotechnical Society Standards, for which we would like to express our heartfelt thanks.

January 2018

Committee for the English Translation of JGS Standards

Preface

It is my great pleasure to present herein the results of the recent international activities of the Japanese Geotechnical Society (JGS). You can find here the first release of the English translation of a variety of JGS technical regulations and more will follow later. These regulations address the fundamental elements of construction practice through establishing a systematic approach from soil investigation to design analysis and performance prediction.

It is well known that the success of construction practice depends on the quality of material tests, field investigation, and design calculation that have been validated through many experiences. Those methodologies are specified by codes. In case that field investigation is carried out in a wrong way, the obtained data does not suit the design calculation. Because of the strong interdependence between practice and code, construction projects should employ both practical technologies and related codes in an assembled body. To facilitate this, JGS is now translating its codes.

As the Vice President for Asia of the International Society for Soil Mechanics and Geotechnical Engineering (ISSMGE), the natural conditions in Asia are very difficult as exemplified by many severe natural disasters and very thick soft clay deposits in river deltas. Typhoons/cyclones provide a huge amount of precipitation in many parts of Asia and increase the amount of river sedimentation. Thus, tens of meters of very soft deposit is encountered. It is very meaningful for Asian engineers to develop a system of construction practice that suits this difficult natural conditions. The first step toward this goal is the development of a new code in which natural conditions in Asia and even past experiences are taken into account. Because the JGS codes have been developed under the natural environment in Asia, they can provide Asian engineers a starting platform for better Asian codes.

Nothing is perfect in this world. For example, many criticisms are made against uncertainties involved in Standard Penetration Tests (SPT). However, SPT provides us with specimens of subsoils for direct inspection. Moreover, many practical formulae rely on correlation with SPT data. Therefore, it is more meaningful to improve the details of SPT practice and stablish a code for SPT rather than discarding SPT as a non-standardized penetration test. Thus, it is important for all practitioners to make efforts to establish new codes. The provided JGS code can be a good starting point for this.

I would cordially invite you all to pay attention to this code to get a better scope on soil mechanics and geotechnical engineering.

Best wishes

Ikuo Towhata
Professor Emeritus of Geotechnical Engineering,
University of Tokyo
Former President of the Japanese Geotechnical Society
Vice President for Asia of ISSMGE

Japanese Geotechnical Society Standards
－Geotechnical and Geoenvironmental Investigation Methods－

1 Procedure for Translating Standards into English

This document contains the geotechnical and geoenvironmental investigation standards of the Japanese Geotechnical Society (JGS). The work of translating the standards into English is scheduled to take about 3 years from 2014. The English language standards are scheduled to be published in 3 separate volumes (Vol. 1 to Vol. 3) in order of priority of those standards most frequently used by practitioners. This is because we want to provide engineers with these Japanese geotechnical and geoenvironmental investigation standards as soon as possible to enable them to understand and refer to Japanese survey equipment together with Japanese geotechnical and geoenvironmental investigation standards. We feel that by translating into English the technical investigation standards that are used by Japanese engineers every day, the standards will be more meaningful for geotechnical and geoenvironmental investigations.

The following tables show all of the geotechnical and geoenvironmental investigation standards produced by JGS. This table of standards shows the order of publication according to the color of the standard. The standards shown with no color will be published in 2018 (Vol. 3). The standards with a shaded background were published in 2015 (Vol. 1), and the standards shown in white characters against a colored background were published in 2016 (Vol. 2). By publishing in English the JGS standards that are most frequently used by engineers, engineers will be able to easily refer to these geotechnical and geoenvironmental investigations standards in the course of their work.

The Japanese Geotechnical Society has cultivated specialist engineers, scholars, and researchers for over more than 60 years. The geotechnical and geoenvironmental investigation standards produced by the Society incorporate much of the know-how that has matured over this long period. By carrying out geotechnical and geoenvironmental investigations, design, and research with these standards by your side, the Japanese Geotechnical Society hopes that you will be able to contribute to the development of engineering and the economic development of your country.

Table 1 List of geotechnical and geoenvironmental investigation standards part 1

- Standards published in the first round (vol.1 in 2015)
- Standards published in the second round (vol.2 in 2016)
- Standards published in the third round (vol.3 in 2018)

Classification	Title	JGS Number	JIS Number	Vol.
Preliminary geotechnical investigations	Method for engineering classification of rock mass	JGS 3811		1
	Method for investigation on geometrical information of discontinuity distribution in rock mass	JGS 3821		1
Geophysical prospecting and well logging	Method for electrical logging	JGS 1121		2
	Method for seismic velocity logging	JGS 1122		1
Soil sampling	Method for obtaining soil samples using thin-walled tube sampler with fixed piston	JGS 1221		1
	Method for obtaining soil samples using rotary double-tube sampler	JGS 1222		1
	Method for obtaining soil samples using rotary triple-tube sampler	JGS 1223		1
	Method for obtaining samples using rotary double-tube sampler with sleeve	JGS 1224		2
	Method for obtaining soil block samples	JGS 1231		2
	Method for obtaining soft rock samples by rotary tube sampling	JGS 3211		2
Groundwater investigations	Method for measuring groundwater level in borehole	JGS 1311		3
	Method for measuring groundwater level in well	JGS 1312		3
	Method for measuring pore water pressure using electric transducer in borehole	JGS 1313		3
	Method for determination of hydraulic properties of aquifer in single borehole	JGS 1314		3
	Method for pumping test	JGS 1315		3
	Method for determination of hydraulic conductivity of compacted fill	JGS 1316		3
	Method for flow layer logging by tracer	JGS 1317		3
	Method of detection of direction and velocity of groundwater flow in single borehole	JGS 1318		3
	Method for determination of hydraulic properties of rock mass using instantaneous head recovery technique in single borehole	JGS 1321		3
	Method for determination of hydraulic conductivity of rock mass using injection technique in single borehole	JGS 1322		3
	Method for Lugeon Test	JGS 1323		3
Sounding	Method for standard penetration test		JIS A1219	1
	Method for mechanical cone penetration test		JIS A1220	1
	Method for Swedish weight sounding test		JIS A1221	1
	Method for field vane shear test	JGS 1411		2
	Method for portable cone penetration test	JGS 1431		2
	Method for portable dynamic cone penetration test	JGS 1433		2
	Method for electric cone penetration test	JGS 1435		2
	Method for dynamic cone penetration test	JGS 1437		3
	Method for soil hardness test	JGS 1441		3
	Method for rebound hammer test on rocks	JGS 3411		3
	Method for point load test on rocks	JGS 3421		3
	Method for needle penetration test	JGS 3431		3
Loading tests	Method for plate load test on soils for road		JIS A1215	2
	Test method for the california bearing ratio (cbr) of in-situ soil		JIS A1222	2
	Method for plate load test	JGS 1521		1
	Pressuremeter test to evaluate index of the ground	JGS 1531		1
	Methods for in-situ direct shear test on rocks	JGS 3511		1
	Pressuremeter test to evaluate mechanical properties of the ground	JGS 3531		1
	Method for borehole jack test	JGS 3532		1
Site density tests	Test method for soil density by the compacted sand replacement method	JGS 1611		3
	Test method for soil density by the water replacement method	JGS 1612		3
	Test method for soil density by the sand replacement method		JIS A1214	2
	Test method for soil density using core cutter	JGS 1613		3
	Test method for soil density using nuclear gauge	JGS 1614		3

Table 2 List of geotechnical and geoenvironmental investigation standards part 2

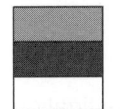

Standards published in the first round (vol.1 in 2015)
Standards published in the second round (vol.2 in 2016)
Standards published in the third round (vol.3 in 2018)

Classification	Title	JGS Number	JIS Number	vol.
Site measurement of soil behaviors	Method for measuring displacement of ground surface using stakes	JGS 1711		3
	Method for measuring settlement of ground surface using settlement plate	JGS 1712		3
	Method for measuring vertical displacement of embankment using cross arm settlement gauge	JGS 1718		3
	Method for measuring tilt of ground surface using tiltmeter	JGS 1721		3
	Method for measuring displacement of the ground surface using extensometer	JGS 1725		2
	Method for measuring ground movement using strain gauge	JGS 1731		2
	Methods for measuring convergence and vertical displacement of crown in underground openings in rocks	JGS 3711		2
	Methods for monitoring rock displacements using borehole extensometers	JGS 3721		2
	Method for monitoring ground displacements using borehole inclinometer	JGS 3722		2
	Method for pull-out test of rock bolts installed in rock mass	JGS 3731		1
	Method for initial stress measurement by overcoring technique using multi-axial strain gauge	JGS 3741		1
	Method for initial stress measurement by compact conical-ended borehole overcoring technique	JGS 3751		1
Method of investigation for soil and groundwater contamination	Method for obtaining samples for environmental chemical analysis using double tube sampler with sleeve	JGS 1911		2
	Method for obtaining samples for environmental chemical analysis using direct push Method	JGS 1912		2
	Method for obtaining subsurface soil samples for environmental chemical analysis	JGS 1921		2
	Method for obtaining groundwater samples for environmental chemical analysis from monitoring well	JGS 1931		2
	Method of soil gas sampling by direct conduction for environmental chemical analysis	JGS 1941		1
	Method of active soil gas sampling for environmental chemical analysis	JGS 1942		1
	Method of passive soil gas sampling for environmental chemical analysis	JGS 1943		1
	Method for air permeability test in vadose zone	JGS 1951		3

2 JGS and JIS Numbers

Each of the geotechnical and geoenvironmental investigation standards in this list is an investigation standard produced by the Japanese Geotechnical Society. Each standard has "JGS" in front of a number. The "JIS" that is written with each of the JGS numbers is an abbreviation for "Japanese Industrial Standard." Japanese Industrial Standards are recognized by the Japanese Ministry of Economy, Trade and Industry of Japan and are based on the JGS standard. In these standards, each investigation standard number is given a JGS number and also a matching JIS number. This means that the standard produced by the Japanese Geotechnical Society is a Japanese standard. This is sufficient for users to understand. In practice, either number may be used.

3 Sponsors of the Translation of the Standards into English

The English language versions of these standards have been produced through donations from Japanese construction companies, design consultants, and testing organizations. The Japanese Geotechnical Society expresses its heartfelt thanks to each company that has made a donation.

An advertisement for each company that has made a donation will be published at the end of the standards. We would be pleased if you would refer to these advertisements when using these geotechnical and geoenvironmental investigations standards and use the advertised products and services in your work.

Geotechnical and Geoenvironmental Investigation Methods - Vol. 3

Table of Contents

1. Groundwater investigation
   - JGS 1311 Method for measuring groundwater level in borehole
   - JGS 1312 Method for measuring groundwater level in well
   - JGS 1313 Method for measuring pore water pressure using electric transducer in borehole
   - JGS 1314 Method for determination of hydraulic properties of aquifer in single borehole
   - JGS 1315 Method for pumping test
   - JGS 1316 Method for determination of hydraulic conductivity of compacted fill
   - JGS 1317 Method for flow layer logging by tracer
   - JGS 1318 Method of detection of direction and velocity of groundwater flow in single borehole
   - JGS 1321 Method for determination of hydraulic properties of rock mass using instantaneous head recovery technique in single borehole
   - JGS 1322 Method for determination of hydraulic conductivity of rock mass using injection technique in single borehole
   - JGS 1323 Method for Lugeon Test

2. Sounding
   - JGS 1437 Method for dynamic cone penetration test
   - JGS 1441 Method for soil hardness test
   - JGS 3411 Method for rebound hammer test on rocks
   - JGS 3421 Method for point load test on rocks
   - JGS 3431 Method for needle penetration test

3. Site density tests
   - JGS 1611 Test method for soil density by the compacted sand replacement method
   - JGS 1612 Test method for soil density by the water replacement method
   - JGS 1613 Test method for soil density using core cutter
   - JGS 1614 Test method for soil density using nuclear gauge

4. Site measurement of soil behaviors
   - JGS 1711 Method for measuring displacement of ground surface using stakes
   - JGS 1712 Method for measuring settlement of ground surface using settlement plate
   - JGS 1718 Method for measuring vertical displacement of embankment using cross arm settlement gauge
   - JGS 1721 Method for measuring tilt of ground surface using tiltmeter

5. Method of investigation for soil and groundwater contamination
   - JGS 1951 Method for air permeability test in vadose zone

Japanese Geotechnical Society Standard (JGS 1311-2012)
Method for measuring groundwater level in borehole

1 Scope

This standard specifies methods for measuring the level of groundwater in sandy and gravelly ground using a borehole. This measurement applies to sandy or gravelly ground near the bottom of the borehole.

Note 1: The objective of this measurement shall not be to measure fluctuations in groundwater level on a long-term or ongoing basis.

Note 2: This measurement shall be used to determine directly the level of groundwater in sandy or gravelly ground with high permeability in situ. High permeability shall be defined as a hydraulic conductivity greater than approximately 10^{-6} m/s.

2 Normative references

None

3 Terms and definitions

The terms and definitions used in this standard are as follows:

3.1 Groundwater level

Groundwater level is the head of pore water in sandy or gravelly ground at a specified depth. In general, this value is converted into and denoted as vertical distance from a reference level. Groundwater level is determined by measuring the equilibrium water level in a borehole.

3.2 Equilibrium water level

Equilibrium water level is the water level when the fluctuations in a water level that has been artificially changed have generally subsided.

4 Equipment

4.1 Drilling equipment

The drilling equipment comprises a boring machine and related equipment that can drill a borehole of a specified diameter down to a specified depth.

4.2 Measurement pipe

The measurement pipe has a specified inner diameter [1] and is made of a material that can withstand the load from being pounded in or press-fit.

Note [1] Specified inner diameter refers to a diameter that will enable drilling within the measurement section and insertion of the water level meter.

Note: A pipe that will not leak from the pipe joint shall be used.

4.3 Water level meter

The water level meter shall enable readings in cm increments.

Note: Fig. A.1 in Annex A shows an example of a water level meter that is used frequently.

5 Preparation of measurement hole

5.1 Drilling down to upper edge of water shielding section

Determine a water shielding section above the measurement section. Then drill a borehole of the specified diameter [2] down to the upper edge of the water shielding section.

Note [2]   Borehole of the specified diameter refers to a diameter that will permit installation of the measurement pipe.

5.2 Water shielding

Install the measurement pipe down to the bottom of the borehole. Then either pound in the measurement pipe from the bottom of the borehole down to the specified depth or press-fit it to provide water shielding from the groundwater in the borehole.

Note 1:   Fig. A.2 in Annex A shows an example of a common method used to conduct water shielding.

Note 2:   If mud fluid or other stabilization liquid has been used when drilling down to the upper edge of the water shielding section, replace the water in the measurement pipe with fresh water.

Note 3:   The pipe used to protect the borehole wall may be installed down to the bottom of the borehole and used for water shielding. The length of the water shielding section should be 50 cm or more.

Note 4:   When measuring confined groundwater level, multiple pipes that use the cohesive soil layer should be used for water shielding from the aquifer above. In such cases, a thorough study shall be made of the effect of water shielding on the permeability of the cohesive soil and the layer thickness.

Note 5:   If water shielding using the measurement pipe is difficult, this should be done in combination with the use of sealing material such as cement milk or water-swelling rubber or water sealing using an air packer.

5.3 Drilling of measurement section

Drill the required measurement section from the end of the measurement pipe using fresh water.

Note 1:   The length of the measurement section should be approximately 50 cm. However, the length shall be determined as needed considering the properties of the soil in the target ground, the layer thickness and so on.

Note 2:   Drilling using fresh water shall be done both to flush the measurement section and to avoid harming the permeability of the measurement section.

6 Measurement method

6.1 Changing of water level in measurement pipe

Change the water level in the measurement pipe temporarily by pumping up water, etc.

Note: Pumping water and injecting water are methods that are available for use in changing the water level in the measurement pipe, but the pumping method should be used.

6.2 Measurement

Measure the water level in the measurement pipe at regular intervals using the water level meter. Continue measurement until the restored water level reaches the equilibrium water level.

Note: The determination as to whether the water level in the measurement pipe has reached equilibrium shall be determined based on the restored water level and the figure showing progress over time. However, in most cases, the water level at which the change in water level over the course of one hour is less than 1 cm is considered to be the equilibrium water level.

7 Processing of test results

A drawing showing the process of measurement water level restoration shall be prepared and used to determine the equilibrium water level, and this shall be used as the groundwater level.

8 Reporting

The following items shall be reported.

a) Location of measurement borehole, ground elevation and top of borehole elevation

b) Depth of measurement section

c) Measurement date and time and weather conditions

d) Type of aquifer (artesian or unconfined) and type of soil

e) Drawing of measurement pipe installation and water shielding method

Note: The inner diameter and outer diameter of the measurement pipe, the inner diameter and outer diameter of the water shielding pipe, the length of the water shielding section, and the water shielding method shall be reported.

f) Water level meter used

g) Restoration curve for water level in measurement pipe

h) Groundwater level

i) If the method used deviates in any way from this standard, give details of the method used.

j) Other reportable matters

Annex A
(Reference)

Example of water level meter and water shielding method

A.1 Example of water level meter

Fig. A.1 shows an example of a water level meter (water surface detection method).

A.2 Example of water shielding method

Fig. A.2 shows an example of a water shielding method.

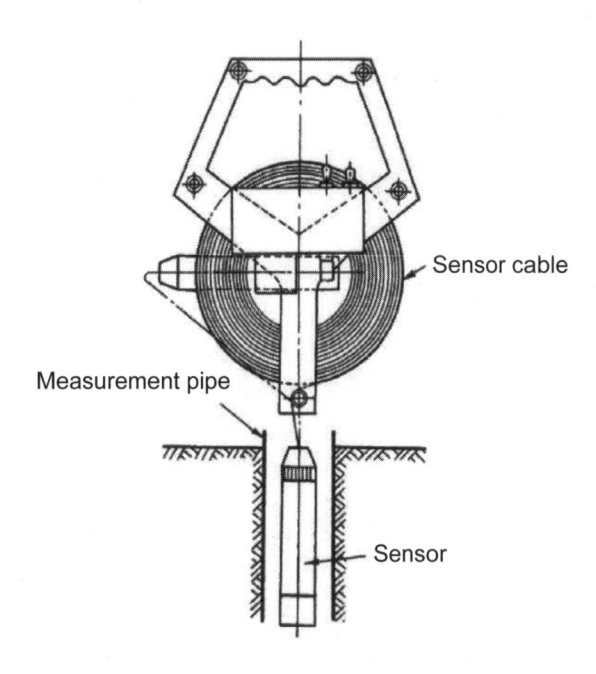

Fig. A.1 Example of water level meter (water surface detection method)

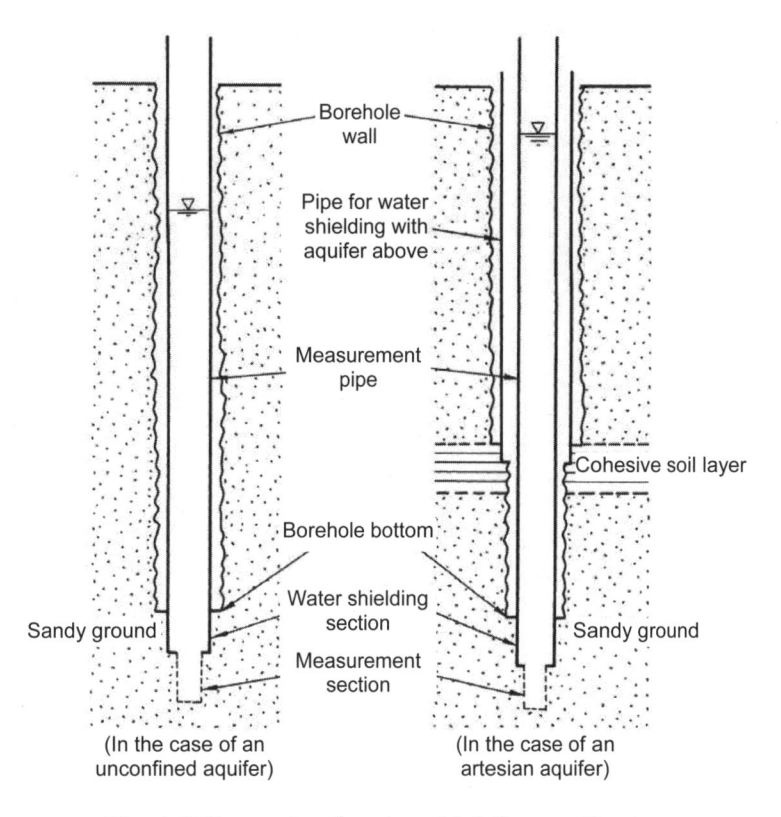

Fig. A.2 Example of water shielding method

Japanese Geotechnical Society Standard (JGS 1312-2012)
Method for measuring groundwater level in well

1 Scope

This standard specifies a method for determining the level of groundwater in an observation well on a long-term basis. This measurement applies to sandy or gravelly ground.

Note: This measurement shall be used to determine directly the groundwater level of sandy or gravelly ground with high permeability in situ. High permeability shall be defined as a hydraulic conductivity greater than approximately 10^{-6} m/s.

2 Normative references

None

3 Terms and definitions

The terms and definitions used in this standard are as follows:

3.1 Groundwater level

Groundwater level is the head of pore water in sandy or gravelly ground at the depth at which a screen is placed. In general, this value is converted into and denoted as vertical distance from a reference level.

4 Equipment

4.1 Drilling equipment

The drilling equipment is a boring machine and related equipment that can drill a borehole of a specified diameter down to a specified depth.

4.2 Observation well

The observation well is created by placing a screen in the target section and inserting a water level measurement pipe into the borehole.

Note 1: In general, the inner diameter of the water level measurement pipe should be approximately 50 mm.

Note 2: The material of the water level measurement pipe and screen (which can also be called a strainer) shall be selected based on consideration of the measurement period, water quality, earth pressure and so on.

Note 3: The opening rate of the screen shall be as large as possible considering the material strength of the water level measurement pipe, in order to enable sufficient water passage capability to be secured. Normally this should be 3 - 10%, or 10% or more in the case of a high material strength. However, if the hole size of the screen is large, a mesh or the like shall be used to protect the hole section so sand does not flow in underground.

4.3 Water level meter

The water level meter shall be capable of measuring the groundwater level on a long-term basis.

Note: The water level meter shall be chosen from among hydraulic type and surface level detection type water level meters (electrode type, float type) (Fig. A.1 in Annex A of JGS 1311, etc.) and so on, in accordance with the purpose of measurement and the observation well conditions.

5 Observation well placement method

The method used to place the observation well shall be as follows.

Note 1: Fig. A.1 in Annex A shows an example of the placement of an observation well.

a) Whenever possible, drilling shall be fresh-water drilling.

Note 2: The diameter of the drilled hole shall be determined based on consideration of the outer diameter of the water level measurement pipe, the filter thickness and so on.

Note 3: Normally, an overbreak should be provided at the base of the drilled hole in order to install the water level measurement pipe at the specified depth.

Note 4: Mud fluid or other stabilization liquid may be used when this is unavoidable in order to protect the borehole wall and so on.

b) Install the specified water level measurement pipe equipped with a screen into the borehole and position it vertically.

Note 5: The screen position and length shall be determined based on the depth and thickness of the target aquifer.

Note 6: When filter material is used, the material shall have a hydraulic conductivity that is greater than the hydraulic conductivity of the ground, and there shall be little inflow of fine soil underground into the observation well.

Note 7: If there is a danger of fine soil flowing in, a grit catcher shall be provided at the bottom end of the screen.

c) Use sealing material or the like to provide a water shield in the gap between the borehole wall and the water level measurement pipe to flow in the vertical direction that may affect the measurement results.

Note 8: Determine the length of the water shield section based on a consideration of the water pressure that will act on the sealing material. Normally it should be 50 cm or more. Cement milk, bentonite, chemical leavening agents or other materials may be used as sealing material. The sealing material shall be determined based on the measurement period, water quality, geological conditions and so on.

Note 9: Take steps to ensure that there is no water leakage from the joint of the water level measurement pipe.

d) Block off the borehole opening side to prevent the infiltration of surface water into the observation well.

e) Flush the observation well thoroughly.

Note 10: If the water level is sufficiently higher than the screen, a pump or bailer or the like should be used to pump up water for flushing. If the water level is near the screen, flushing should be accomplished through repeated injecting and pumping. Be sure to flush particularly carefully if mud fluid or other stabilization liquid has been used.

6 Measurement method

The measurement method shall be as follows.

a) Insert the water level meter into the observation well.

b) Measure the water level at specified intervals.

Note 1: Exclude the initial data, as it will be affected by flushing and the installation of the water level meter.

Note 2: When measurements are conducted over a long period of time, be sure to flush the inside of the screen as needed.

Note 3: Set the interval for water level measurement in accordance with the purpose of the investigation. If the water level varies because it is affected by the tidal fluctuations in locations near the coast, by a well that is operating, or by rainfall and so on, set the measurement interval so the effect of such factors can be evaluated.

Note 4: When using an automatic water level meter, conduct verification periodically by using a manual water level meter or the like.

7 Processing of test results

The relationship between the measured water level and elapsed time shall be shown in a table or depicted in graphic form.

8 Reporting

The following items shall be reported.

a) Number and location of observation well, ground elevation and elevation of top of pipe

b) Depth of measurement section

Note 1: For the depth of the measurement section, the depth from the surface of the ground to the bottom of the sealing material and the depth down to the bottom of the hole, as shown in Fig. A.1 in Annex A, shall be reported.

c) Measurement date and time

Note 2: The weather conditions, air pressure, rainfall and other conditions at the time of water level measurement shall be reported if necessary.

d) Structural diagram of observation well

e) Water level meter used

f) Results of water level measurement

g) If the method used deviates in any way from this standard, give details of the method used.

h) Other reportable matters

Annex A
(Reference)

Example of observation well

A.1 Example of observation well

Fig. A.1 shows an example of an observation well.

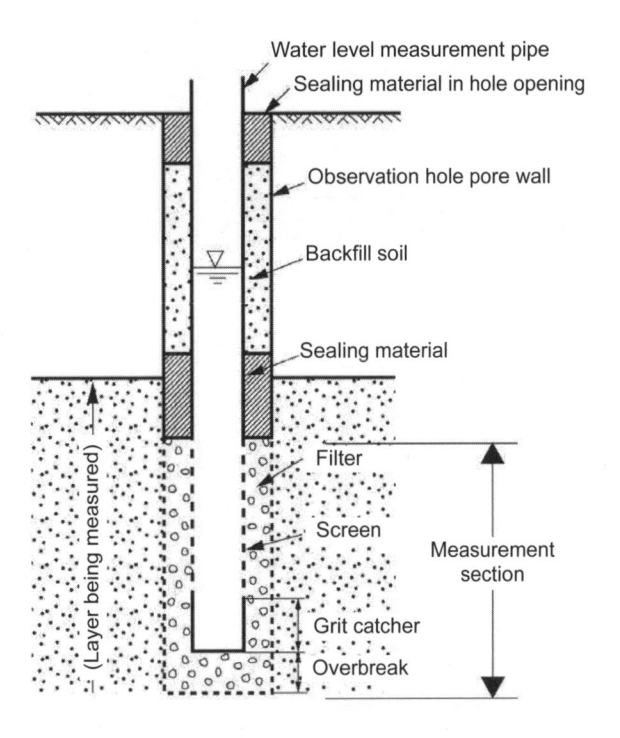

Fig. A.1 Example of observation well

Japanese Geotechnical Society Standard (JGS 1313-2012)
Method for measuring pore water pressure using electric transducer in borehole

1 Scope

This standard specifies a method for determining pore water pressure in situ using an electric transducer installed in a borehole. This measurement applies to saturated ground.

Note: This measurement shall be used to determine directly the pore water pressure in saturated sandy ground or cohesive soil ground in situ.

2 Normative references

None

3 Terms and definitions

The terms and definitions used in this standard are as follows:

3.1 Pore water pressure

Pore water pressure is the pressure of the water filling the pore space in the soil in situ.

3.2 Electric transducer

The electric transducer shall be able to convert the pore water pressure into an electrical signal.

3.3 Equilibrium water pressure

Equilibrium water pressure is the pore water pressure when the fluctuations in the pore water pressure following the installation of the electric transducer have generally stabilized.

4 Equipment

4.1 Drilling equipment

The drilling equipment is a boring machine and related equipment that can drill a borehole of a specified diameter down to a specified depth.

4.2 Electric transducer and cable

The electric transducer is a pressure transducer that renders it capable of measuring water pressure only, and it shall be able to convert from electrical signals to water pressure values. An electric transducer with the capacity needed for the purpose of measurement and the anticipated range of the pore water pressure and having an accuracy of ±1% of the maximum capacity shall be selected. A configuration with a filter or the like in the pressure receiver shall be used. The cable shall be capable of transmitting the electrical signal to the readout unit.

Note: Fig. A.1 in Annex A shows an example of an electric transducer.

4.3 Readout unit

The readout unit shall be capable of displaying the electrical signal or water pressure values that have been converted by the electric transducer.

Note: The readout unit shall be compatible with the electric transducer.

5 Electric transducer installation method

5.1 Preparations

The following preparations shall be made for installation of the electric transducer.

a) Remove any slime from the bottom of the borehole.

b) At ground level, saturate the filter in the pressure receiver with water.

Note: Deaerated water should be used to saturate the filter in the pressure receiver.

c) Connect the cable from the electric transducer to the readout unit.

5.2 Installation of electric transducer

Methods for installation of the electric transducer include the push-in method and the backfill method. To confirm that the electric transducer and readout unit are functioning properly, record the reading on the readout unit during the installation of the electric transducer. In the event that the push-in installation method is used to push the transducer into the ground, check to make sure there is no overload.

5.2.1 Installation of electric transducer using the push-in method

While successively tightening the electric transducer unit and the rod, slowly install the electric transducer into the borehole. When the electric transducer reaches the bottom of the borehole, push it in gently to the specified depth.

Note: Be careful not to apply excessive force to the electric transducer. Also, note that if the electric transducer is not pushed in far enough, a water shield will not be created. As a general guide, the electric transducer shall be pushed in to a depth of 30 cm or more. Fig. A.2 (a) in Annex A shows an example of the installation of the electric transducer.

5.2.2 Installation of electric transducer using the backfill method

Lower the electric transducer carefully to the specified depth and then backfill by adding sand or other filter material. Then use bentonite or other sealing material to create a secure water shield.

Note: Check the placement depth by marking scale gradations or the like on the cable. Fig. A.2 (b) in Annex A shows an example of the installation of the electric transducer.

6 Measurement method

6.1 Measurement prior to installation

In 5.1 c), record the reading D_0 on the readout unit in the no-load state at ground level.

Note 1: Take the reading D_0 in the no-load state and confirm the value noted in the calibration table for the electric transducer.

Note 2: During installation, the changes in the reading D on the electric transducer depending on the depth from the groundwater table should be recorded, and then confirm the calibration coefficient A and the reading D_0 in the no-load state.

6.2 Measurement of equilibrium water pressure

Record the reading D on the readout unit and the elapsed time following installation, and then measure the amount of time until the reading reaches equilibrium. Record the reading D when an equilibrium state is reached, and use that value to calculate the equilibrium water pressure (initial value).

Note 1: The equilibrium water pressure is the water pressure when the excess pore water pressure produced during installation has dissipated and the changes in water pressure have stabilized.

Note 2: When the hourly change in the readout value is less than the readout accuracy of the electric transducer, equilibrium is considered to have been achieved.

6.3 Measurement of long-term pore water pressure

When conducting long-term measurements, check the equilibrium pressure in 6.2 and then record the reading D on the readout unit at specified time intervals.

7 Processing of results

The results shall be processed as follows.

a) The pore water pressure shall be determined using the following equation:

$$\rho_W = A(D - D_0)$$

where

ρ_W : Pore water pressure (kPa)
A : Calibration constant (kPa / readout value unit)
D : Reading
D_0 : Reading in no-load state

b) The relationship between the measured pore water pressure and measurement time shall be depicted in graphic form.

8 Reporting

The following items shall be reported.

a) Number and location of measurement borehole, and ground elevation

b) Measurement depth

c) Measurement date and time

Note 1: The weather conditions, atmospheric air pressure, temperature, rainfall and other conditions at the time of water pressure measurement shall be reported if necessary.

d) Specification[1], dimensions, and calibration table[2] of the electric transducer used,

Note [1] Electric transducer specification refers to model, capacity and sensitivity. Dimensions refers to shape and size.

Note [2] The calibration table shall show the relationship between pressure and readout value.

e) Method used for installation of electric transducer

Note 2: The filter deaeration method, the installation method (push-in or backfill) and the water shielding method shall be clearly indicated.

f) Pore water pressure values and curve showing changes over time in pore water pressure

g) If the method used deviates in any way from this standard, give details of the method used.

h) Other reportable matters

Annex A
(Reference)

Example of electric transducer and electric transducer Installation

A.1 Example of electric transducer

Fig. A.1 shows an example of an electric transducer.

A.2 Example of electric transducer installation

Fig. A.2 shows an example of the installation of an electric transducer.

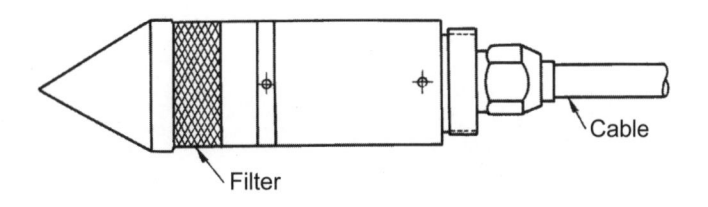

Fig. A.1 Example of electric transducer

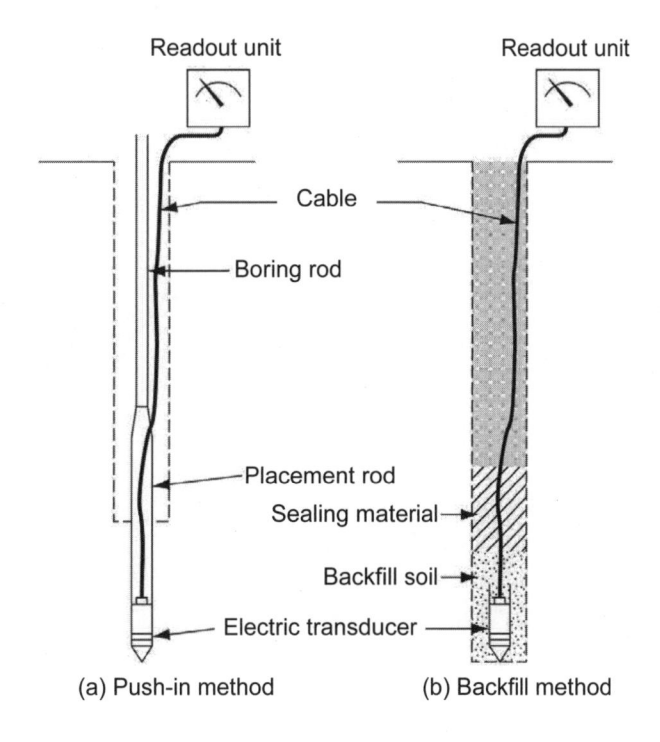

(a) Push-in method (b) Backfill method

Fig. A.2 Example of electric transducer installation

Japanese Geotechnical Society Standard (JGS 1314-2012)
Method for determination of hydraulic properties of aquifer in single borehole

1 Scope

This standard specifies methods for determining the hydraulic conductivity of ground using a single borehole or a single well. This test applies to saturated ground beneath the level of groundwater.

2 Normative references

The following standards shall constitute a part of this standard by virtue of being referenced in this standard. The latest versions of these standards shall apply (including supplements).

JGS 1311   Method for measuring groundwater level in borehole
JGS 1312   Method for measuring groundwater level in well

3 Types of test method

The following two methods are available for use as test methods. The method should be chosen based on the permeability of the ground.

3.1 Transient method

The water level in the measurement pipe is temporarily raised or lowered and the changes over time in the water level as it returns to an equilibrium state are measured in order to determine the hydraulic conductivity of the ground.

Note 1: Care shall be taken when using the transient method to measure sandy ground or gravelly ground whose hydraulic conductivity is expected to be about 10^{-4} m/s or more, as the changes in water level over time occur rapidly and will be difficult to measure.

Note 2: The transient method is suitable for use when the water level recovers approximately 90% of the initial water level differential during the period of the test, and when at least 10 measurement data points can be obtained during that time.

3.2 Steady-state method

Water is withdrawn or injected in and the flow rate when the water level in the measurement pipe becomes constant is measured in order to determine the hydraulic conductivity of the ground.

Note 1: The steady-state method is suitable for sandy or gravelly ground when the hydraulic conductivity is expected to be approximately 10^{-5} m/s or more.

Note 2: The steady-state method is suitable for application when water level recovery is rapid and 10 or more valid measurement data points cannot be obtained with the transient method at the specified time intervals.

4 Equipment

4.1 Common items

4.1.1 Drilling equipment

The drilling equipment is a boring machine and related equipment that can drill a borehole of a specified diameter down to a specified depth.

4.1.2 Measurement pipe

The measurement pipe shall be as specified in JGS 1311 and JGS 1312.

Note 1: The inner diameter of the measurement pipe shall be the same over the water level fluctuation range.

Note 2: When a well is used, a well casing shall be treated as the measurement pipe.

4.1.3 Water level meter

The water level meter shall be able to conduct measurements over time of the water level in the measurement pipe, and shall be capable of readings in cm increments.

Note 1: A water pressure gauge may be used to measure the water level.

Note 2: Prior to using the water pressure gauge, the calibration coefficient and the indicator value in no-load status shall be checked and the gauge shall be calibrated.

4.2 In the case of the transient method

4.2.1 Water pumping equipment

The water pumping equipment shall be able to temporarily lower the water level in the measurement pipe in a comparatively short period of time.

Note: For the transient method, a pump or bailer shall be used to lower the water level in the measurement pipe. Equipment to make the interior of the measurement pipe airtight and lower the water level using air pressure may also be used. Removing the slag that has been submerged in the measurement pipe will enable the water level in the measurement pipe to be lowered temporarily.

4.2.2 Water injecting equipment

The water injecting equipment shall be able to temporarily raise the water level in the measurement pipe in a comparatively short period of time.

Note: The water level in the measurement pipe can be raised by injecting water from a bucket or a water tank or the like, or by adding slag.

4.3 In the case of the steady-state method

4.3.1 Water pumping equipment

The water pumping equipment shall be able to temporarily pump up water from inside the measurement pipe at a constant flow rate.

Note: A pump that is capable of pumping water continuously at a constant flow rate shall be used to lower the water level in the measurement pipe.

4.3.2 Water injecting equipment

The water injecting equipment shall be able to add water into the measurement pipe at a constant flow rate.

Note: A pump that is capable of adding water continuously at a constant flow rate shall be used to raise the water level in the measurement pipe.

4.3.3 Flow meter

The flow meter shall be able to measure the water pumping flow rate or water injection flow rate.

4.3.4 Water storage tank

The water storage tank shall store the groundwater that has been pumped up so it does not return to the ground. In the case of water injection, the water storage tank shall be used to store the water for injection.

5 Preparation of test borehole

The test borehole shall be prepared as follows. If a well is used, the method used to prepare the well shall be in accordance with JGS 1312.

a) In accordance with JGS 1311, drill down to a location above the test section and conduct water shielding. Then install the measurement pipe.

Note 1: If there is a danger that the borehole wall may collapse, use a casing to protect the borehole wall.

b) Drill the required test section from the bottom of the measurement pipe. Then use fresh water to flush the test section thoroughly prior to the start of the test.

Note 2: If there is a danger that a section of the test section may collapse, place a measurement pipe with a screen at that section.

c) The L/D ratio between the length L of the test section and the borehole diameter D of the test section shall be 4 or greater. The shape of the test section shall be kept constant until the end of the test.

Note 3: If an L/D ratio of $L/D \geq 4$ cannot be secured, evaluate the test results using an equation that differs from the equations shown in A.1 and A.3 in Annex A.

6 Test method

Note: Fig. B.1 in Annex B shows an example of a test method.

6.1 Measurement of equilibrium water level

Before starting the test, measure the equilibrium water level in the test section in accordance with JGS 1311.

Note: When a water pressure gauge is used, the water pressure measurement shall be converted into a water level value.

6.2 Testing with the transient method

The test procedure to be used in the case of the transient method shall be as follows.

a) Temporarily lower the water level in the measurement pipe by pumping water, or temporarily raise the water level by injecting water.

Note 1: When injecting water, do so carefully to avoid clogging in the test section or allowing air bubbles to enter.

b) Measure the water level h (m) in the measurement pipe over time. The water level shall be measured in cm and the time shall be measured in seconds.

Note 2: The time at which no more fluctuations are caused in the water level in the measurement pipe shall be the test start time ($t = 0_s$). If the water level is being measured continuously with a water pressure gauge beginning prior to the test, the point at which the fluctuations in the water level in the pipe are at their highest value shall be the test start time ($t = 0_s$), and the water level at that time shall be water level at the time of test commencement h_p (m).

Note 3: The water level shall be measured based on the ground height of the borehole or the depth of the water pressure gauge as a reference. Choose a point that will not move during the test as the reference point.

6.3 Testing with the steady state method

The test method to be used in the case of the steady state method shall be as follows.

a) Pump water from inside the measurement pipe or inject water to the measurement pipe.

Note: When injecting water, do so carefully to avoid clogging in the test section or allowing air bubbles to enter.

b) Measure the water level h (m) in the measurement pipe over time.

c) Measure the pumping flow rate or injecting flow rate Q_0 (m³/s) when the water level in the measurement pipe has become constant.

7 Processing of results

In the case of the transient method, prepare a water level recovery curve based on the measurements of the water level in the measurement pipe. In the case of the steady-state method, record the pump-up flow rate or injection flow rate and the steady-state value of the water level in the measurement pipe. Annex A shows a method for use in analyzing the results.

8 Reporting

The following items shall be reported.

a) Number and location of test borehole, and ground elevation

Note 1: The altitude should be determined and used as the ground elevation.

b) Depth of test section (before and after test)

c) Test date and time and weather conditions

d) Structure of test borehole

e) Method used to measure water level

f) Location of reference point for water level measurement

Note 2: The borehole aperture height or water pressure gauge installation depth shall be reported as the reference point for water level measurement.

g) Method used to measure flow rate

h) Test method

i) Equilibrium water level in test section

j) Water level measurement records

k) Method used to evaluate results in the case of the transient method and corresponding water level recovery curve

Note 3: A $\log_{10} s - t$ curve or $s/s_P - \log_{10} t$ curve shall be displayed.

l) In the case of the steady-state method, the pump-up flow rate or injection flow rate and the water level in the measurement pipe

m) Hydraulic conductivity

Note 4: In the case of the transient method, if the test results have been evaluated using the type curve matching method, the specific storage coefficient S_s (1/m) shall be reported.

n) If the method used deviates in any way from this standard, give details of the method used.

o) Other reportable matters

Annex A
(Regulation)

Methods for analyzing results

A.1 Transient method: Straight line method

The straight line method shall be used when the test results are judged to be not affected by the storage effect of the ground.

The straight line method shall be used when a straight line section is determined to be present in the $\log_{10} s - t$ curve shown in Fig. B.2 in Annex B.

The straight line method can determine only the hydraulic conductivity.

The method used to process the results obtained using the straight line method shall be as follows.

a) Plot the difference in water level $s = |h_0 - h|$ (m) between the equilibrium water level h_0 (m) and the water level in the measurement pipe h (m) on the logarithmic scale (vertical axis) of a semi-log graph, and plot the time t (s) on the arithmetic scale (horizontal axis) of the graph, in order to create the $\log_{10} s - t$ curve shown in Fig. B.2 in Annex B. Check to see if a straight line can be discerned in the plotted graph.

b) Determine the gradient of the straight line section $(1/s)$. The gradient can be determined from two arbitrary points $(t_1, \log_{10} s_1)$ and $(t_2, \log_{10} s_2)$ on the line using the following equation.

$$a = \frac{\log_{10}(s_1 / s_2)}{t_2 - t_1}$$

c) Calculate the hydraulic conductivity k (m/s) using the following equation.

$$k = \frac{(2.3 d_e)^2}{8L} \log_{10}\left(\frac{2L}{D}\right) a$$

for

$$\frac{L}{D} \geq 4$$

where

d_e: $d_e = d$ in the case of a manual water level meter

When a hydraulic pressure meter is used, the diameter of a circle with an area equal to the equivalent cross-sectional area determined by subtracting the cross-sectional area of the water level measurement cable c (m^2) from the cross-sectional area of the interior of the measurement pipe.

$$\left(= \sqrt{d^2 - \frac{4c}{\pi}}\right) \text{ (m)}$$

d: Inner diameter of measurement pipe in water level fluctuation section (m)

D: Borehole diameter of test section or outer diameter of measurement pipe screen (m)
L: Length of test section (m)

A.2 Transient method: Type curve matching method

The type curve matching method shall be used when the test results are judged to be affected by the storage effect of the ground and there is no clear linear section in the $\log s - t$ curve.

The curve matching method can determine the hydraulic conductivity and the specific storage coefficient.

Note: As the estimation sensitivity of the specific storage coefficient is significantly lower than that of the hydraulic conductivity, it is used as a reference value.

The method used to process the results obtained with the type curve matching method shall be as follows.

a) Determine the difference in water level during the test $s = |h_0 - h|$ (m) and the difference in water level at the start of the test $s_P = |h_0 - h_P|$ (m) from the equilibrium water level h_0 (m), the water level measured during the test h (m), and the water level at the start of the test h_P (m). Also determine the water level differential ratio s/s_P.

b) Plot the water level differential ratio s/s_P on the arithmetic scale (vertical axis) of a semi-log graph and plot the elapsed time since the start of the test t (s) on the logarithmic scale (horizontal axis) of the graph to plot the measurement values.

c) On a different semi-log graph with the same scale as the graph you created in b), create a group of standard curves showing the relationship between the water level differential ratio s/s_P for each storage coefficient ratio α and the dimensionless time β, as shown in Fig. B.3 in Annex B.

d) As shown in Fig. B.4 in Annex B, overlay the two graphs that you created in b) and c). Move one graph in parallel to the direction of the time axis (horizontal axis) and select the standard curve that most closely matches the measurement value. Read the value for α that corresponds to this type curve (α_m) and the time axis coordinates t_m and β_m for both graphs that correspond to the arbitrary matching points.

e) Using the following equations, determine the hydraulic conductivity k (m/s) and the specific storage coefficient S_S (1/m).

$$k = \frac{d_e^2 \beta_m}{4 L t_m}$$

$$S_S = \frac{d_e^2}{L D^2} \alpha_m$$

A.3 Steady-state method

Use the following equation to determine the hydraulic conductivity k (m/s) from the amount of fluctuation s_0 (m) in the water level in the measurement pipe from the equilibrium water level in steady-state condition and the flow rate at steady-state status Q_0 (m³/s).

$$k = \frac{Q_0}{2\pi s_0 L} \ln\left(\frac{2L}{D}\right) = \frac{2.3 Q_0}{2\pi s_0 L} \log_{10}\left(\frac{2L}{D}\right)$$

for

$$\frac{L}{D} \geq 4$$

where

Q_0 : Pumping flow rate or fill flow rate (m³/s)

s_0 : Amount of fluctuation in water level in steady-state condition (m)

Annex B
(Reference)

Example of analyzing results

B.1 Example of test method

Fig. B.1 shows an overview of the test method.

B.2 Example of straight line method (transient method)

Fig. B.2 shows an example of the straight line method.

B.3 Example of type curve matching method (transient method)

Fig. B.3 and Fig. B.4 show examples of the type curve matching method.

(a) Transient method

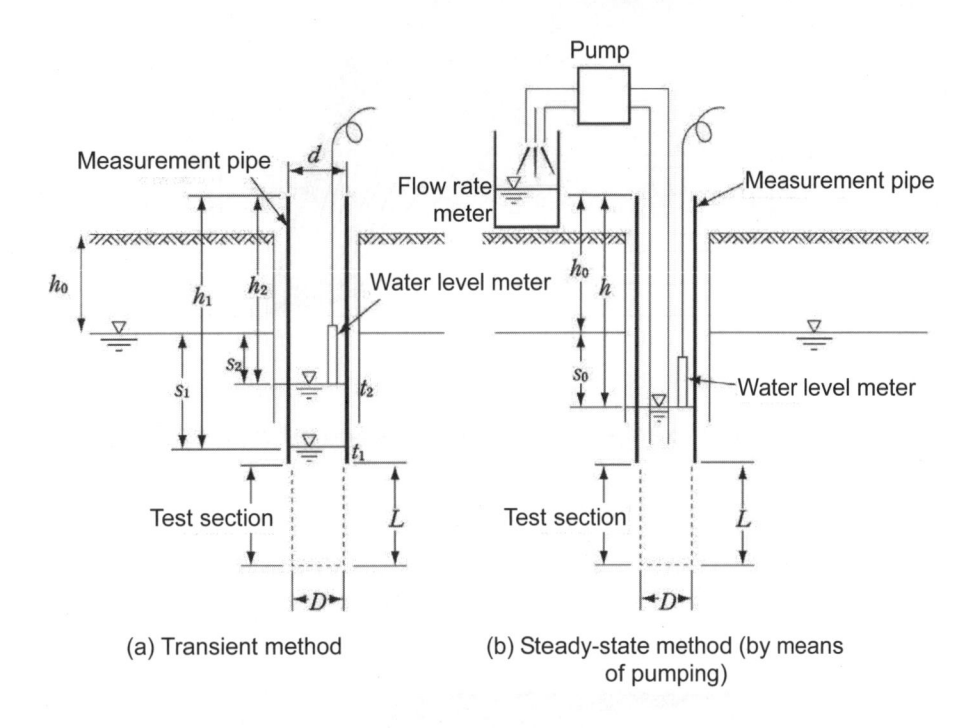

(b) Steady-state method (by means of pumping)

Fig. B.1 Overview of the test method

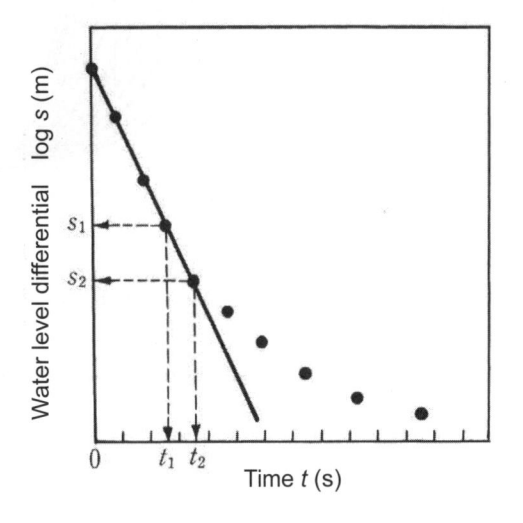

Fig. B.2 Example of the $\log_{10} s - t$ curve

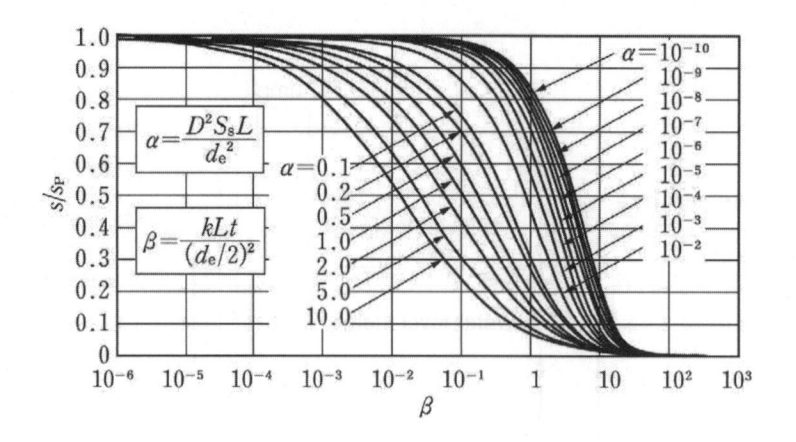

Fig. B.3 Example of the group of standard curves

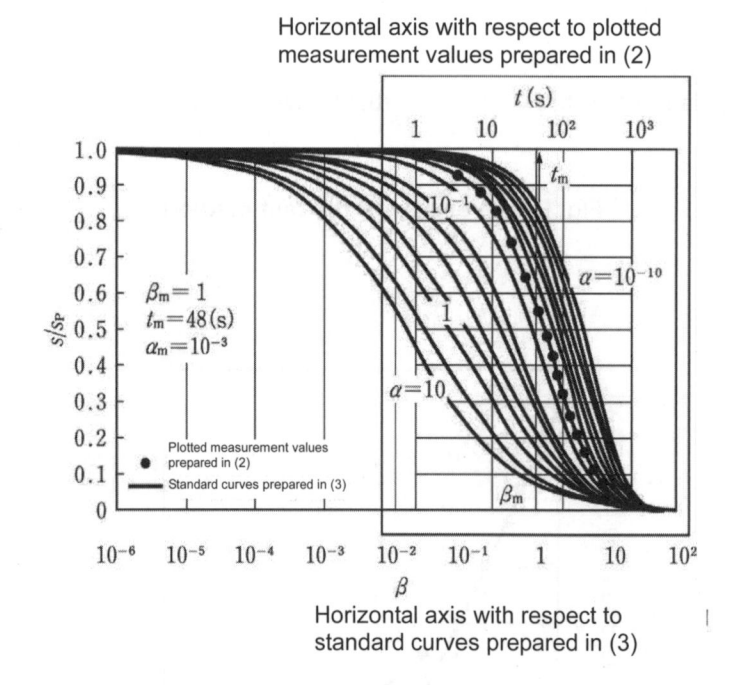

where $\alpha = \dfrac{D^2 S_s L}{d_e^2}$, $\beta = \dfrac{kLt}{(d_e/2)^2}$

Fig. B.4 Example of the curve matching method

$\beta_m = 1$
$t_m = 48\,(s)$
$\alpha_m = 10^{-3}$

Japanese Geotechnical Society Standard (JGS 1315-2012) Method for pumping test

1 Scope

This standard specifies methods for determining the transmissivity (or hydraulic conductivity) and storage coefficient of an aquifer. This test applies to saturated aquifers.

Note: The test applies to ground in which the fluctuations in groundwater level in an observation well during pumping can be observed. As a general guide, the test applies to ground whose hydraulic conductivity is greater than 10^{-6} m/s.

2 Normative references

The following standards shall constitute a part of this standard by virtue of being referenced in this standard. The latest versions of these standards shall apply (including supplements).

JIS B 8302   Measurement methods of pump discharge
JGS 1312   Method for measuring groundwater level in well
JGS 1313   Method for measuring pore water pressure using electric transducer in borehole

3 Terms and definitions

The main terms and definitions used in this standard are as follows:

3.1 Pumping test

The pumping test is a method for determination of hydraulic properties in situ using a pumping well and multiple observation wells. The drop in the water level in the pumping well and observation wells during pumping and the recovery of the water level after pumping is stopped are measured over time to determine the transmissibility (or hydraulic conductivity) and storage coefficient of the aquifer.

3.2 Pumping well

The pumping well is a well that is used to pump up ground water during the pumping test, in order to cause a drop in the level of groundwater in the ground that is being tested.

3.3 Observation well

This is a general term for wells and boreholes that make it possible to measure the fluctuations in the level of groundwater during the pumping test. It includes pore water pressure measurement boreholes in which a water pressure gauge has been installed.

3.4 Aquifer

An aquifer is permeable formation such as a sand layer or gravel layer that stores groundwater.

4 Equipment

4.1 Pump

The pump shall have the capacity to control the specified pumping flow rate [1].

Note [1]   The specified flow rate is the pumping flow rate that is sufficient to create fluctuations in the groundwater level that can be observed in the observation wells.

Note 1: The capacity of the pump is determined after estimating the pumping flow rate based on the results of a geological survey and existing references and so on, and a lifting pump that can control a constant pumping flow rate is selected.

Note 2: When executing the recovery test, a check valve is placed in the middle of the pipe (pumping pipe) from the lifting pump to prevent the water in the pumping pipe from flowing back into the pumping well.

4.2 Pumping flow rate meter

The pumping flow rate meter shall be the triangular weir specified in JIS B 8302 "Measurement methods of pump discharge," or a flow meter, container whose capacity is known or other apparatus that is capable of measuring the pumping flow rate of the pump.

4.3 Water level meter

The water level meter shall be the water level meter specified in JIS 1312 or the electric transducer specified in JGS 1313.

5 Installation of pumping well and observation wells

The pumping well and observation wells shall be installed as follows.

Note: Fig. B.1 in Annex B shows an example of the installation of the test equipment.

5.1 Pumping well

The pumping well shall be installed as follows.

a) Drill a borehole of the specified diameter [1] and the specified depth [2].

Note [1]: The specified borehole diameter shall be the well diameter that is determined based on the permeability of the ground, the outer diameter of the pipe with screen, the filter thickness and so on.

Note [2]: The specified drilling depth shall be the depth down to the bottom end of the aquifer that is the target of the test. However, in order to place the pipe with screen at the specified depth, normally an overbreak shall be provided at the bottom of the drilled borehole.

Note 1: The method shall be fresh water drilling, but mud fluid or other stabilization fluid may be used when this is unavoidable in order to protect the borehole wall and so on.

Note 2: A rotary type, percussion type or other type of drilling machine shall be used in accordance with the type of soil, drilling depth, drilling diameter and so on.

b) Install the pipe with screen vertically into the borehole.

Note 3: The position and length of the screen shall be determined based on the depth and thickness of the target aquifer. The water withdrawal section of the pumping well shall be the section between the screen and the ground that has been filled with filter material.

Note 4: The pipe with screen shall have an inner diameter that enables the lifting pump to be installed.

Note 5: The material of the pipe and the screen shall be selected based on a consideration of the measurement period, water quality, soil pressure and so on.

Note 6: The opening rate and shape of the screen shall be such that groundwater is not prevented from flowing into the pumping well but also ensures that as little as possible of the filter material and the fine soil in the ground flows in.

Note 7: A sand reservoir measuring 1 - 2 m in length should be provided at the bottom end of the pumping well.

c) Fill the gap between the borehole wall at the screen placement depth and the pipe with filter material.

Note 8: A filter material shall be selected that has a hydraulic conductivity that is sufficiently higher than that of the ground, and that prevents to the greatest extent possible the inflow of fine soil in the ground into the pumping well.

d) Provide water shielding for the gap between the borehole wall and the pipe at depths other than the screen placement depth, in order to prevent the inflow of groundwater from strata other than the aquifer that is the target of the test. The water shielding method shall be in accordance with JGS 1312.

Note 9: A water shielding method and water shielding material such that collapse due to erosion caused by the flow of water from pumping will not occur shall be selected.

e) Flush the pumping well thoroughly.

Note 10: Pumping, swapping, jetting, air lift or other flushing method shall be selected as appropriate in accordance with the pumping well structure and the drilling method.

Note 11: Flush particularly carefully if mud fluid or other stabilization liquid has been used to protect the borehole wall.

5.2 Observation wells

The observation wells shall be installed as follows.

a) The method that is used to install the observation wells shall be in accordance with JGS 1312.

Note 1: The screen placement depth in the observation wells should be the same as the depth of the water withdrawal section in the pumping well.

Note 2: Observation wells shall be placed densely near the pumping well and less densely as the distance from the pumping well increases. Three or more observation wells should be placed on a single measurement line.

The locations for placement of observation wells shall be at generally equally spaced sections when the distance from the pumping well is plotted on a logarithmic scale.

The observation wells should be placed on two orthogonal measuring lines centered on the pumping well.

b) The placement method when a pore water pressure meter is used shall be in accordance with JGS 1313.

Note 3: In the case of a pore water pressure meter, the meter should be placed near the center depth of the pumping well water withdrawal section.

6 Test method

6.1 Test preparations

The following test preparations shall be made.

a) Measure the top end of the pumping well that will serve as the reference for water level measurement and the top end of each observation well, or the elevation at the location where the pore water pressure gauge is placed, and survey the distance γ from the center of the pumping well to the center of each observation well.

b) Install the pump and water level meter inside the pumping well and install the water level meters inside the observation wells.

c) To avoid the effect of fluctuations in groundwater level due to the installation of instrument inside the wells and the flushing process and so on, measure the status of water level fluctuations and confirm that the fluctuations have generally subsided.

d) To determine the natural fluctuations in the level of the groundwater, conduct observations in advance for a period of time equivalent to the testing period and determine the equilibrium water level h_0.

Note: In the event that the target groundwater fluctuates because it is affected by tides, air pressure or human groundwater use or the like, determine the relationship between groundwater level and tides, air pressure etc. and the effect of human groundwater use and so on in advance, and correct the observation water level. Determine the equilibrium water level from the water level after correction.

6.2 Step draw down test

The step draw down test method shall be as follows.

a) Change the pumping flow rate at determined intervals and measure the relationship between the pumping flow rate Q and the water level in the pumping well h_w during each interval.

Note: Take steps to ensure that the pumped water will not affect the groundwater level in the target aquifer.

b) Process the results in a) as the relationship between the pumping flow rate Q and the drawdown in the pumping well s_w.

c) After the step draw down test, stop pumping and continuously observe the groundwater level until it can be confirmed that the level has been generally restored.

6.3 Constant flow rate pumping test

The constant flow rate pumping test method shall be as follows.

a) Set the target pumping flow rate for the constant flow rate pumping test based on the step draw down test results that were processed in 6.2 b).

Note 1: If the relationship between the pumping flow rate Q and the drawdown in the pumping well s_w in the step draw down test is plotted as shown in Fig. B.2 in Annex B, it may show sudden ups and downs. Check the pumping flow rate at which the water level in the pumping well begins to drop suddenly (limit pumping flow rate Q_c) or the pumping flow rate at which continuous inflow of sand into the pumping well occurs, and set the target pumping flow rate for the constant flow pumping test in a range below these values. However, make sure that significant groundwater level fluctuations in the observation wells can be measured.

b) Begin pumping at the target pumping flow rate determined in 6.3 a), and measure the elapsed time after pumping was started t and the water level h in the pumping well and the observation wells over time.

Also measure the pumping flow rate Q_p to confirm that water is being pumped at a constant flow rate.

Note 2: Take steps to ensure that the pumped water will not affect the groundwater level in the target aquifer.

Note 3: The rate of decrease in the groundwater level due to pumping will diminish over time at an exponential rate, so set the measurement time intervals so the measurement times are at generally equivalent intervals when viewed on a logarithmic scale.

c) When the water level in the pumping well and the observation wells has become generally constant, maintain that status (steady state status) for several hours, and then measure the pumping flow rate Q_p and the water level h_c in the pumping well and the observation wells.

Note 4: Depending on the objective of the test, the steady state status may not need to be checked.

6.4 Recovery test

After the constant flow rate pumping test, stop pumping and measure the elapsed time after pumping was stopped t' and the water level h in the pumping well and each observation well. The test is complete when the water level in each observation well has been generally restored to the equilibrium water level.

Note: The rate of recovery of the groundwater level due to pumping being stopped will diminish over time at an exponential rate, so set the measurement time intervals so the measurement times are at generally equivalent intervals when viewed on a logarithmic scale.

7 Analysis of test results

The decrease in water levels and the amount of recovery of the water levels in the pumping well and observation wells, and the pumping flow rate over time, shall be measured during the test, and the results shall be analyzed.

The method of analysis of the results that is generally used is shown in Annex A.

Note: During the test, the test results should be plotted on a graph to determine the fluctuation status.

8 Reporting

The following items shall be reported.

a) Numbers and locations of pumping well and observation wells

b) Ground elevation and elevation of top end of pumping well and observation wells

Note 1: If a pore pressure meter has been installed, report the elevation of the location at which the pore pressure meter has been installed

c) Type of aquifer (confined or unconfined) and type of soil

d) Structural drawing of pumping well and observation wells

e) Water level measurement method

f) Method used to measure pumping flow rate

g) Test date and time

Note 2: The weather conditions, air pressure, rainfall, tidal fluctuations and so on at the time of the test shall be reported if necessary.

h) Results of observations conducted prior to the test

i) Results of step draw down test (diagram showing the relationship between the pumping flow rate and the decrease in the water level in the pumping well)

j) Results of constant flow rate pumping test (pumping flow rate and changes over time in the water level in the pumping well and observation wells)

k) Results of recovery test (time elapsed after pumping was stopped and changes over time in the water level in the pumping well and observation wells)

l) Transmissibility, hydraulic conductivity, and storage coefficient of aquifer (specific storage in the case of a confined aquifer and specific yield in the case of an unconfined aquifer), as well as the radius of influence and the method used to determine this value

m) If the method used deviates in any way from this standard, give details of the method used.

n) Other reportable matters

Annex A
(Regulation)

Method used to analyze results

A.1 Common items

The common items relating to the method used to analyze the results are as follows.

a) The method used to analyze results shown here is based on the theoretical formulas that are derived under the conditions satisfied by the following hypotheses. It may be applied when the conditions under which the test was executed do not deviate significantly from these conditions.

   1) Aquifer is in confined condition.

   2) Aquifer has an infinite horizontal expanse.

   3) Aquifer is homogeneously isotropic and has a uniform thickness.

   4) Groundwater level prior to test is horizontal.

   5) Water is pumped at a constant flow rate.

   6) Flow of groundwater in aquifer during pumping is horizontal in a radial fashion.

b) Methods of organization A.2, A.3 and A.5 may be used when, in addition to the above, the following conditions have been met.

   7) Water in aquifer flows instantly along the gradient of the hydraulic head.

   8) Diameter of pumping well is sufficiently small and the effect of storage in the pumping well is negligible.

Note 1: If the decrease in water level during the test is small in the pumping test for an unconfined aquifer, the above methods of organization may be applied mutatis mutandis.

Note 2: When using organization methods A.2, A.3 and A.5 to organize test results during pumping and recovery under transient conditions, the results may be organized individually for each observation well, but it is advantageous to organize the observation results for multiple observation wells on the same graph. When conducting tests under ideal conditions, organizing the results in this manner will make it possible to overlay on the same graph the plotted decrease in water level in the observation wells with different distances γ from the pumping well, enabling the validity of the test to be confirmed.

c) The appropriate method of analysis from among those listed below shall be chosen in accordance with the conditions under which the test was executed and the objective of the test.

A.2 Type curve matching method (Theis method)

The type curve matching method is a method for comparing the results of water level measurements made under transient conditions with the theoretical decrease in water level (type curve), in order to determine the transmissibility (or hydraulic conductivity) and the storage coefficient.

The method used to analyze the results obtained from the curve matching method shall be as follows.

Note 1: When using the type curve matching method to determine the transmissibility (or hydraulic conductivity) and the storage coefficient, the matching point on the plotted test data and the type curve may be any arbitrary point selected on the graph. Fig. B.4 in Annex B shows an example of the method used to determine the coordinates for the matching point $[(1/\lambda)_m, W(\lambda)_m], [(t/\gamma^2)_m, s_m]$.

Note 2: The time axis may be a combination of γ^2/t and λ. In such cases, the time coordinates for the matching point shall be $(\lambda)_m$ and $(\gamma^2/t)_m$, and the inverses of these values $(1/\lambda)_m$ and $(t/\gamma^2)_m$ should be substituted in equation e).

a) Calculate the decrease in groundwater level s in the observation well at elapsed time t following the start of pumping, using the following equation.

$$s = h_0 - h$$

where

h_0: Equilibrium water level (m)

h: Water level in observation well at elapsed time t (s) following the start of pumping (m)

b) On a log-log graph, plot the measurement results for each observation well, with the decrease in groundwater level of the observation wells s on the vertical axis and t/γ^2 on the horizontal axis. Here γ is the distance from the pumping well to the observation wells.

c) On a separate log-log graph with the same scale as the one on which the $s - t/\gamma^2$ relationship was plotted, depict the well function $W(\lambda) - (1/\lambda)$ relationship as shown in Fig. B.3 in Annex B as a standard curve.

d) Overlay the two graphs and move them in parallel upward, downward, to the left and to the right and read the coordinates $[(1/\lambda)_m, W(\lambda)_m]$, $[(t/\gamma^2)_m, s_m]$ for an arbitrary point on the graph when the test results and the type curve most closely match one another.

e) Substitute these coordinate values in the following equation to calculate the transmissibility T (or hydraulic conductivity k) and storage coefficient S.

$$T = \frac{Q_P}{4\pi s_m} W(\lambda)_m$$

$$k = \frac{T}{b}$$

$$S = 4T \frac{(t/\gamma^2)_m}{(1/\lambda)_m}$$

$$S_s = \frac{S}{b} \quad \text{(in the case of a confined aquifer)}$$

$$S_y = S \quad \text{(in the case of an unconfined aquifer)}$$

where

Q_P: Pumping flow rate during constant flow pumping test (m³/s)

b: Thickness of aquifer (m) (in the case of an unconfined aquifer $b = h_0$: thickness of saturated aquifer at equilibrium water level)

S_s: Specific storage coefficient (1/m)

S_y: Specific yield

A.3 Straight line method using $s - \log_{10}(t/\gamma^2)$ plot (Jacob's method)

The straight line method using the $s - \log_{10}(t/\gamma^2)$ plot is a method in which the results of water level measurement in the transient condition is plotted onto a semi-log graph, and the transmissibility (or hydraulic conductivity) and storage coefficient are determined from the interval of the plot that is approximated linearly.

The method used for analysis of the results in the straight line method using the $s - \log_{10}(t/\gamma^2)$ plot is as follows.

Note: Fig. B.5 in Annex B shows the method used to determine the gradient a and $(t/\gamma^2)_{S=0}$ when determining the transmissibility (or hydraulic conductivity) and storage coefficient using the straight line $(s - \log_{10}(t/\gamma^2))$ method.

a) Calculate the decrease in groundwater level in the observation well at elapsed time t following the start of pumping, using the following equation.

$$s = h_0 - h$$

b) On a semi-log graph, plot the measurement results for each observation well, with the decrease in groundwater level of the observation wells s on the arithmetic scale (vertical axis) and t/γ^2 on the logarithmic scale (horizontal axis).

c) Select the linearly approximated interval of the $s - \log_{10}(t/\gamma^2)$ plotted data and read the gradient of the straight line (the difference in the s value corresponding to a single logarithmic cycle of the horizontal axis) a and the coordinates $((t/\gamma^2)_{S=0}, 0)$ for the intersection between the straight line extension and the horizontal axis $s = 0$ axis.

d) Calculate the transmissibility T (or hydraulic conductivity k) and storage coefficient S, using the following equations

$$T = \frac{2.3 Q_P}{4\pi a}$$

$$k = \frac{T}{b}$$

$$S = 2.25 T (t/\gamma^2)_{S=0}$$

$$S_s = \frac{S}{b} \quad \text{(in the case of a confined aquifer)}$$

e) Substitute the obtained parameters in the following equation to confirm the validity of the straight line approximation section.

$$\frac{1}{\lambda} = \frac{4T}{S}\left(\frac{t}{\gamma^2}\right)^* \geq 100$$

where

$(t/\gamma^2)^*$: t/γ^2 with respect to straight line approximation section data (s/m²)

If the above equation is not satisfied, perform linear approximation for longer duration test results and return to c).

A.4 Straight line method using $s - \log_{10} \gamma$ plot (Thiem method)

The straight line method using the $s - \log_{10} \gamma$ plot is a method in which the relationship between the distance γ from the pumping well to the observation well and the decrease in water level s_c in the steady state condition at each observation well is plotted on a semi-log graph and the transmissibility (or hydraulic conductivity) is determined from the interval of the plotted data that can be approximated by a straight line.

This method of analysis can be used for test results with condition 2) of the hypotheses shown in A.1 as indicated below.

2) A fixed water level boundary is present at finite radius R in the horizontal direction.

The method used to analyze the results of the straight line method using the $s - \log_{10} \gamma$ plot shall be as follows.

Note: Fig. B.6 in Annex B shows the methods used to determine the gradient a and the radius of influence R when determining the transmissibility (or hydraulic conductivity) using the straight line $(s - \log_{10} \gamma)$ method.

a) Calculate the decrease in groundwater level s_c for each observation well in the steady state condition, using the following equation.

$$s_c = h_0 - h_c$$

where

h_c : Water level in observation well in the steady state condition (m)

b) On a semi-log graph, plot the decrease in steady state groundwater level s_c on the arithmetic scale (vertical axis) and the distance γ from the pumping well to each observation well on the logarithmic scale (horizontal axis) to determine the relationship between these values.

c) Approximate the linear interval of the $s_c - \log_{10} \gamma$ plotted data and read the gradient a (the difference in the s_c value corresponding to a semi-log cycle of the horizontal axis) of this linear interval.

d) Calculate the transmissibility T (or hydraulic conductivity k), using the following equation

$$T = \frac{2.3 Q_P}{2\pi a}$$

$$k = \frac{T}{b}$$

e) Extend the approximated straight line to determine the coordinates for the intersection with the horizontal axis ($s = 0$ axis) and use this distance as the radius of influence R.

A.5 Method using a recovery equation

The recovery equation method is a method in which the results of water level measurement when the level has recovered are plotted on a semi-log graph and the transmissibility (or hydraulic conductivity) is determined from the interval that is approximated by a straight line.

This method of organization can be used for test results with condition 5) of the hypotheses shown in A.1 as indicated below.

5) Water is pumped at a constant flow rate and then pumping is stopped.

The method used to analyze the results obtained from the recovery method shall be as follows.

Note 1: Fig. B.7 in Annex B shows the method used to determine the linear gradient a when using a recovery equation to determine the transmissibility (or hydraulic conductivity).

Note 2: When pumping was stopped and recovery performed at the stage at which the water level decrease behavior during pumping could be linearly approximated on the $s - \log_{10}(t/\gamma^2)$ plot shown in 5.2, the $s=0$ horizontal axis intercept of the approximated straight line is converged to $\log_{10}(t/t')=1$. In contrast, when the water level recovery is performed after pumping has been stopped following the achievement of steady state status, the $s=0$ axis intercept is not converged to $\log_{10}(t/t')=1$. In either case, however, the linear gradient a does not change, so the transmissibility T is not affected.

a) Calculate the decrease in groundwater level s at elapsed time t' after pumping has been stopped, using the following equation.

$$s = h_0 - h$$

where

h : Water level in observation well at elapsed time t' (s) after pumping has been stopped (m)

h_0 : Equilibrium water level (m)

b) On a semi-log graph, plot the decrease in groundwater level of the observation well s on the arithmetic scale (vertical axis) and t/t' on the logarithmic scale (horizontal axis) to determine the relationship between these values for each observation well. Here t is the elapsed time since pumping was started.

c) Select an interval of the $s - \log_{10}(t/t')$ plotted data where a straight line is approximated and read the gradient a (the difference in the s value corresponding to a semi-log cycle of the horizontal axis) for that straight line.

d) Calculate the transmissibility T (or hydraulic conductivity k), using the following equation.

$$T = \frac{2.3 Q_P}{4\pi a}$$

$$k = \frac{T}{b}$$

where

Q_P : Pumping flow rate during constant flow pumping test (m$_3$/s)

Annex B
(Reference)

Examples of pumping test equipment and organization of pumping test results

B.1 Example of test method

Fig. B.1 shows an example of the installation of the pumping test facility.

B.2 Example of step draw down test results

Fig. B.2 shows an example of the step draw down test results.

B.3 Example of type curve

Fig. B.3 shows an example of the type curve.

B.4 Example of method used to determine matching point coordinates in type curve matching method

Fig. B.4 shows an example of the method used to determine the matching point coordinates in the type curve matching method.

B.5 Example of pumping test results using s $s - \log_{10}(t/\gamma^2)$ plot

Fig. B.5 shows an example of the pumping test results using an $s - \log_{10}(t/\gamma^2)$ plot.

B.6 Example of pumping test results using s $s - \log_{10}\gamma$ plot

Fig. B.6 shows an example of the pumping test results using an $s - \log_{10}\gamma$ plot.

B.7 Example of pumping test results using recovery equation

Fig. B.7 shows an example of the pumping test results using a recovery equation.

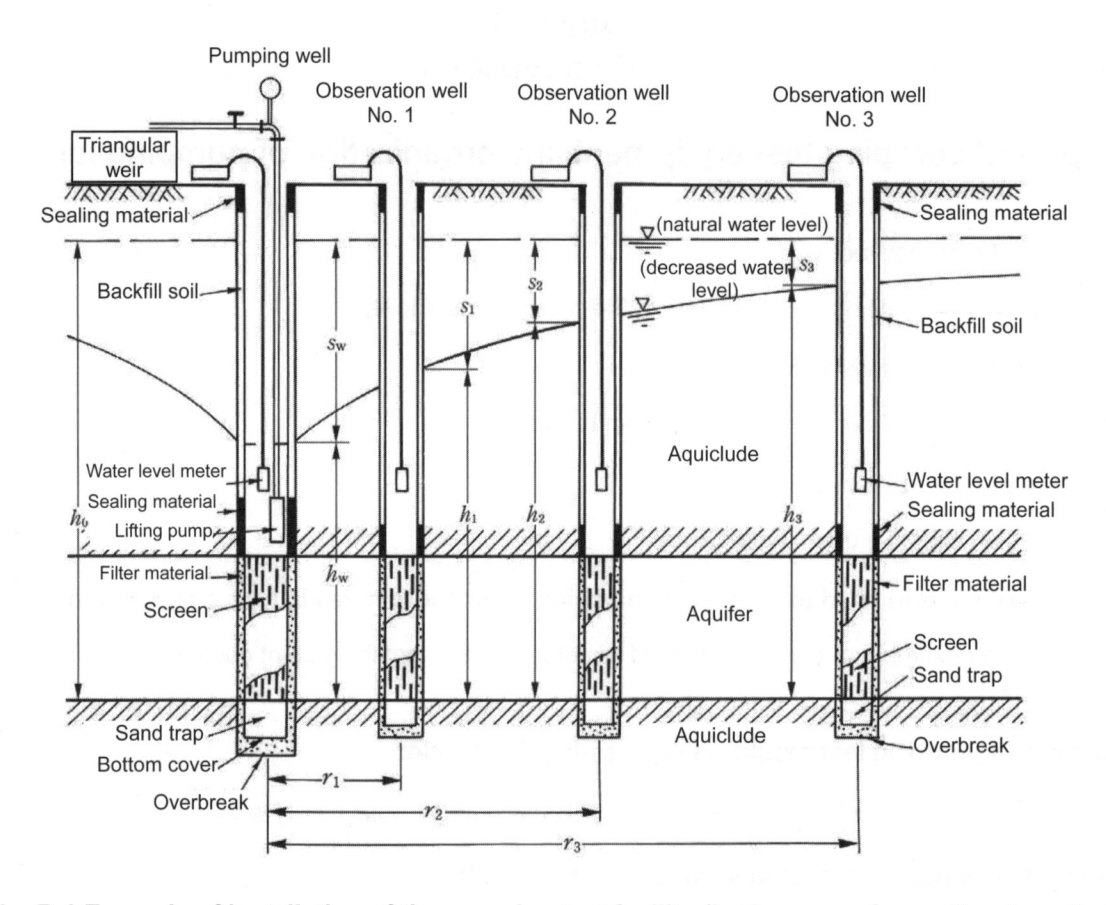

Fig. B.1 Example of installation of the pumping test facility (in the case of a confined aquifer)

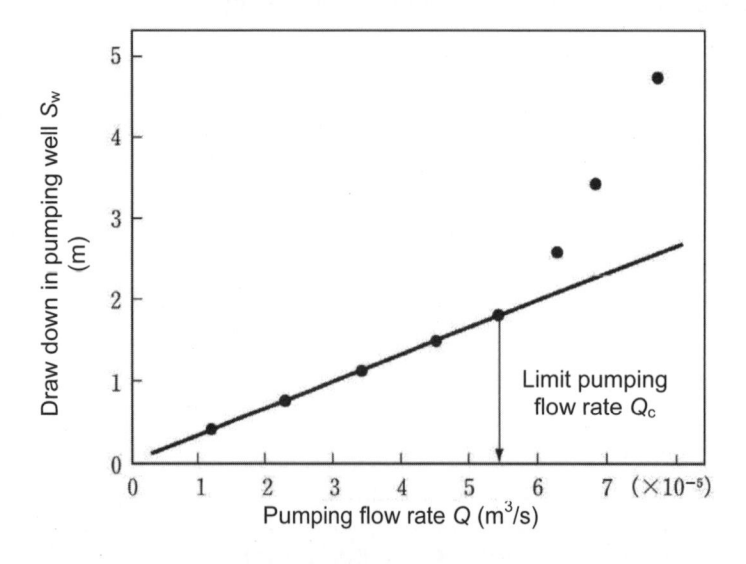

Fig. B.2 Example of the step draw down test results

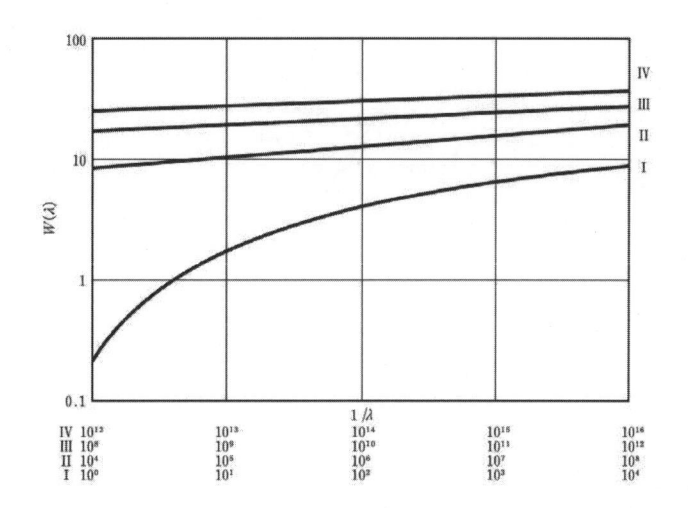

Fig. B.3 Example of the type curve   Well function $W(\lambda)-(1-\lambda)$ curve

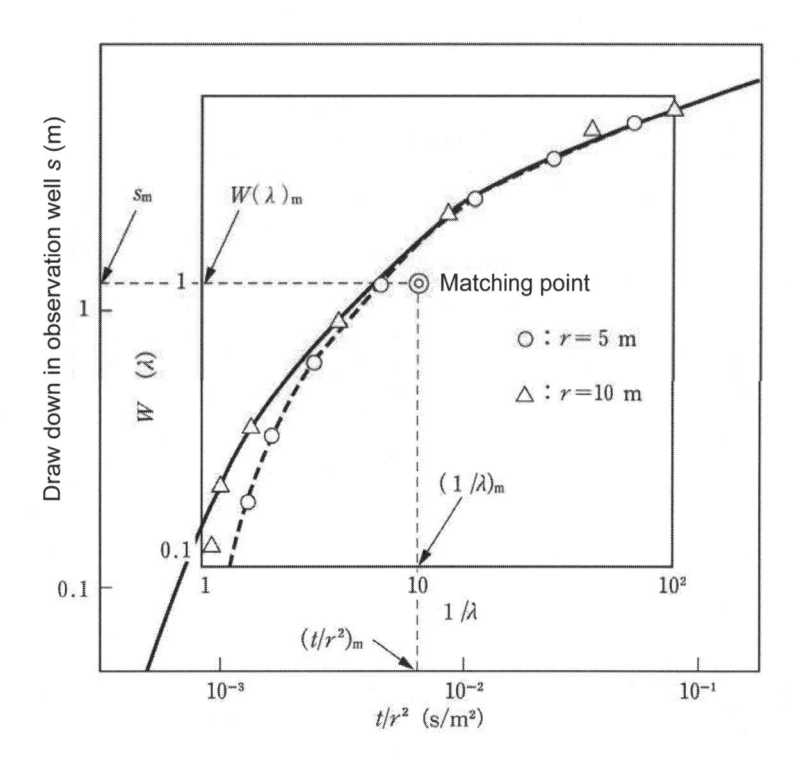

Fig. B.4 Example of method used to determine matching point coordinates in the curve matching method

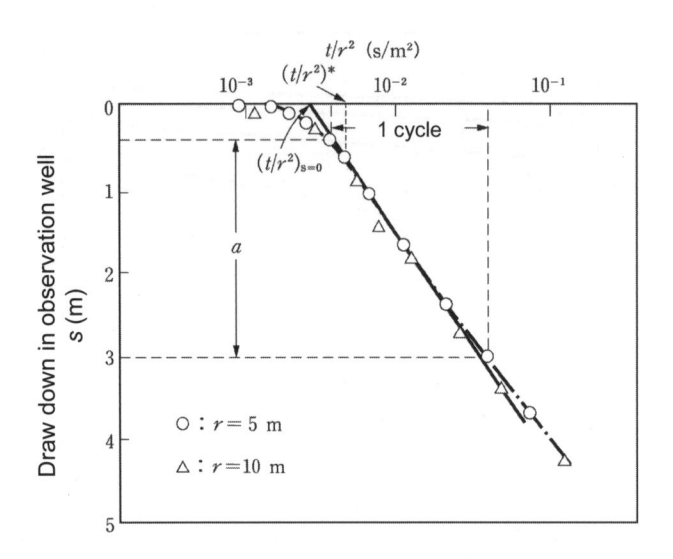

Fig. B.5 Example of pumping test results using $s - \log_{10}(t/\gamma^2)$ plot

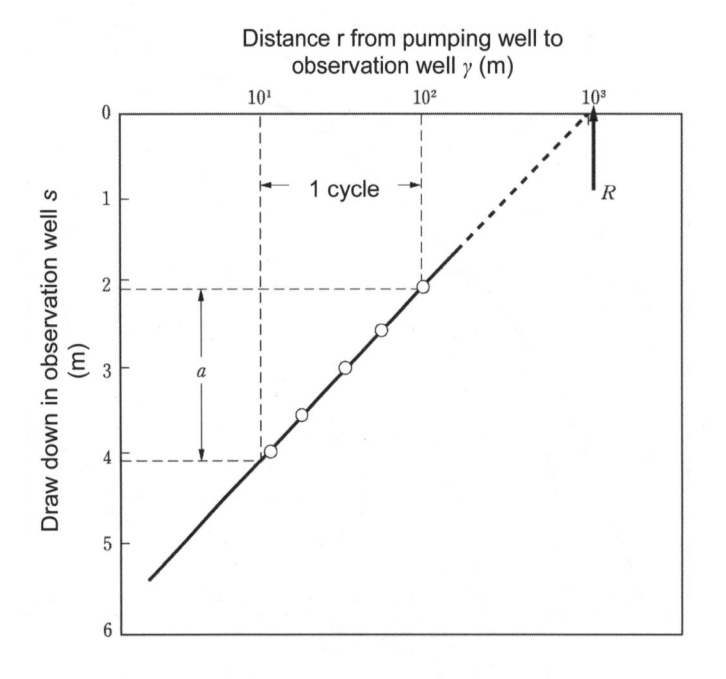

Fig. B.6 Example of pumping test results using $s - \log_{10}\gamma$ plot

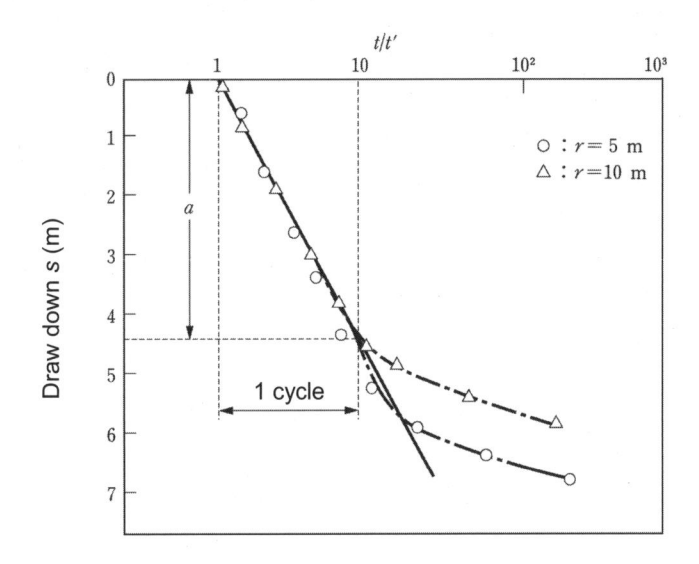

Fig. B.7 Example of pumping test results using recovery equation

Japanese Geotechnical Society Standard (JGS 1316-2012)
Method for determination of hydraulic conductivity of compacted fill

1 Scope

This standard specifies methods for determining the hydraulic conductivity of compacted ground using the steady state method.

This test applies to compacted ground above the groundwater level.

2 Normative references

The following standard shall constitute a part of this standard by virtue of being referenced herein. The latest version of this standard shall apply (including supplements).

JIS A 5005   Crushed stone and manufactured sand for concrete

3 Terms and definitions

The terms and definitions used in this standard are as follows:

3.1 Steady state method

The steady state method is a method for measuring hydraulic conductivity under the condition that the water level in a test hole and rate of the water flow passing through the ground are constant.

3.2 Compacted ground

Compacted ground is ground formed by compacting under approximately optimum water content. It can be applied to dam bodies, dam impermeable soil walls, and so on.

3.3 Marriotte siphon

A marriotte siphon is an apparatus consisting of an airtight water tank having pipes filled with air at an upper location and with water at a bottom location, that enables measurement of the rate of the water flow by reading the water level change in the tank. The apparatus is capable of keeping the water head constant by the location of the end of the upper pipe connected to the air (constant water level maintenance pipe).

A standing pipe type marriotte siphon is also available.

4 Equipment

4.1 Digging equipment

The digging equipment shall consist of a scoop, garden trowel, straight knife etc.

4.2 Crushed stone

The crushed stone shall be equivalent to crushed stone 1505 (particle size 20 - 5 mm) specified in JIS A 5005.

Note: The crushed stone shall be cleanly washed stone.

4.3 Water injection apparatus

The water injection apparatus shall be a Marriotte siphon or other apparatus with equivalent functions (an apparatus capable of measuring the flow rate at a constant water level).

Note 1: Fig. B.1 in Annex B shows an example of a hydraulic conductivity test using a Marriotte siphon.

Note 2: The capacity of the airtight water tank shall be determined taking into consideration the hydraulic conductivity of the ground being measured. For example, if the hydraulic conductivity is approximately 1.0×10^{-8} m/s, the water level h in the test hole is 0.25 m and the test hole radius r_0 is 0.15 m, the steady state flow rate is estimated to be approximately 2.0×10^{-9} m^3/s.

5 Preparations of test

The test shall be prepared as follows.

a) Excavate a cylindrical test hole with a diameter of 0.3 m and a depth of 0.3 m approximately in the ground, to satisfy the relationship of Tu>3h, where h is the depth of the water in the test hole, and Tu is the depth from the groundwater level to the water level in the test hole (see Fig. B.1 in Annex B).

b) Carve the inner wall of the test hole using a straight knife.

c) Install the filling water pipe into the test hole and fill the hole with crushed stone.

d) Open the water supply valve to fill the airtight water tank with water. Then close the valve when the required quantity of water has been filled in the tank.

6 Test method

The test method shall be as follows.

a) Using a bucket, fill the test hole with water. Then open the injection valve to supply water to the test hole through the injection pipe to maintain the water level constant by connecting the end of the constant water level maintenance pipe. Measure the depth h_1 (m) from the ground surface to the water level in the test hole, and calculate the water depth in the test hole h (m) from the depth of the test hole z (m).

b) Read the time t (s) and the water level in the airtight water tank H (m) from the water level gauge.

Note: The measurement intervals shall be set in accordance with the speed of the changes in water level.

c) Repeat b) until the change in the water level in the airtight water tank H per unit time has become constant.

7 Analysis of test results

During the test, the water level in the airtight water tank over time shall be measured and the results shall be processed.

A method for analyzing the results is shown in Annex A.

8 Reporting

The following items shall be reported.

a) Number and location of test borehole, and ground elevation

Note 1: The elevation should be determined as the ground height.

b) Test date and time and weather conditions

Note 2: The water temperature shall be reported if necessary.

c) Test apparatus

Note 3: The diameter and depth of the test hole and the dimensions of the Marriotte siphon shall be reported.

d) Records of measurements of water level in airtight water tank H and time t

e) Hydraulic conductivity

f) If the method used deviates in any way from this standard, give details of the method used.

g) Other reportable matters

Note 4: The ground compaction conditions, degree of saturation, and particle size distribution shall be reported if necessary.

Annex A
(Regulation)

Method used to analyze results

A.1 Analysis of test results

Plot the measurement values with the water level in the airtight water tank H as the vertical axis and the time t as the horizontal axis to produce a $H-t$ diagram. Draw a line in the straight section of the $H-t$ relation in the diagram, and determine the water level measurement times t_1 and t_2.

A.2 Calculation of hydraulic conductivity

The hydraulic conductivity shall be calculated as follows.

a) Calculate the steady state flow rate Q (m₃/s) using the following equation.

$$Q = \frac{A(H_1 - H_2)}{t_2 - t_1}$$

where

    A: Cross-sectional area of airtight water tank ($= \pi a^2$) (m²)
    a: Radius of inner airtight water tank (m)
    t_1, t_2: Times at measurement of water level in airtight water tank (s)
    H_1, H_2: Water level in airtight water tank at times t_1 and t_2 (m)

b) Calculate the hydraulic conductivity k (m/s) using the following equation.

$$k = \frac{Q}{2\pi h^2}\left[\log_e\left[\frac{h}{r_0} + \left\{\left(\frac{h}{r_0}\right)^2 + 1\right\}^{1/2}\right] - \left\{\left(\frac{r_0}{h}\right)^2 + 1\right\}^{1/2} + \left(\frac{r_0}{h}\right)\right]$$

The following condition shall be satisfied:

$T_u > 3h$

where

    Q: Steady state flow rate (m³/s)
    h: Water depth in test hole (m)
    r_0: Radius of test hole (m)
    T_u: Depth from water level in test hole to groundwater level (m)

Annex B
(Reference)

Example of hydraulic conductivity test using Marriotte siphon

B.1 Example of hydraulic conductivity test using Marriotte siphon

Fig. B.1 shows an example of a hydraulic conductivity test using a Marriotte siphon.

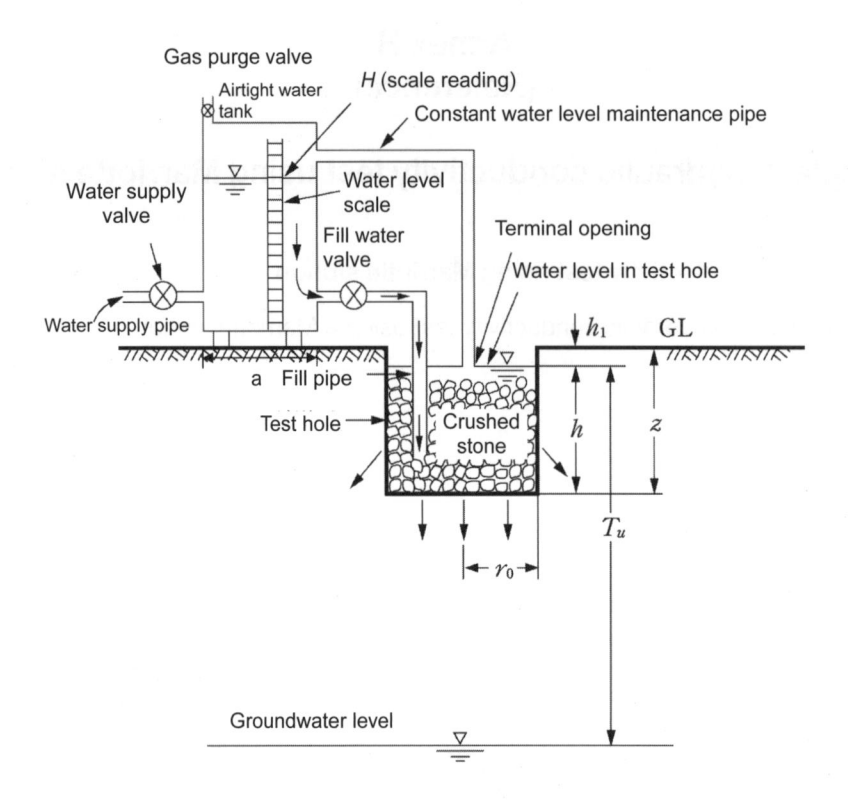

Fig. B.1 Example of hydraulic conductivity test using Marriotte siphon

Japanese Geotechnical Society Standard (JGS 1317-2012) Method for flow layer logging by tracer

1 Scope

This standard specifies methods for detecting the groundwater flow layer in the ground using a single borehole. This test applies to saturated ground beneath the level of the groundwater.

2 Normative references

The following standard shall constitute a part of this standard by virtue of being referenced herein. This standard shall apply the latest version of the following standard (including supplements).

JGS 1311 Method for measuring groundwater level in borehole

3 Terms and definitions

The main terms and definitions used in this standard are as follows:

3.1 Method for flow layer logging by tracer

In this method, a tracer is injected into the borehole and stirred thoroughly to change the electrical resistance or temperature of the groundwater inside the borehole, and then the changes over time in the electrical resistance or temperature of the groundwater in the borehole are measured in order to detect the groundwater flow layer.

3.2 Tracer

A tracer is water with electrical resistance or temperature that is different from the groundwater.

3.3 Groundwater flow layer

The groundwater flow layer is a section of the ground that is highly permeable and through which the groundwater flows relatively quickly.

4 Types of logging method

The following logging methods are available.

a) Logging by means of electrical resistance measurement [1]

Note [1] "Logging by means of electrical resistance measurement" is a general term that includes measurement of the electrical resistance and the resistivity of the water.

b) Logging by means of temperature measurement

5 Equipment

5.1 Drilling equipment

The drilling equipment shall comprise a boring machine and related equipment that can drill a borehole of a specified diameter down to a specified depth.

5.2 Logging pipe

The logging pipe shall have a specified inner diameter [2] and shall be provided with screens at the logging interval.

Note [2]: The specified inner diameter shall be sufficient to enable the water level meter and logging cable to be installed, and shall be approximately 50 mm.

Note 1: Normally, the opening rate of the screen shall be approximately 5 - 10%.

Note 2: A logging pipe is not needed if the borehole wall is self-supported.

5.3 Cleaning equipment

The cleaning equipment shall be used to clean the borehole for the logging.

Note: The cleaning equipment may be a bailer, an air lift device, a pump or other equipment.

5.4 Water level meter

The water level meter shall be the water level meter specified in JGS 1311.

5.5 Logging cable

5.5.1 Logging by means of electrical resistance measurement

The logging cable shall be equipped with one or more electrode pairs.

5.5.2 Logging by means of temperature measurement

The logging cable shall be equipped with temperature sensors at intervals of 25 - 50 cm.

5.6 Measuring instruments

5.6.1 Logging by means of electrical resistance measurement

The instrument shall be capable of measuring the electrical resistance or the resistivity to two or more significant digits.

5.6.2 Logging by means of temperature measurement

The instrument shall be capable of measuring the temperature to 0.1 °C accuracy.

5.7 Tracer

5.7.1 Logging by means of electrical resistance measurement

The tracer shall be water with an electrical resistance that is different from that of the groundwater.

Note: In general, a saline solution is used as the tracer. However, in areas near the coast or other areas in which the groundwater has a high salt concentration, water with a high electrical resistance may be used as the tracer.

5.7.2 Logging by means of temperature measurement

The tracer shall be water with a temperature that is different from that of the groundwater.

Note: In general, hot water is used as the tracer. However, in areas in which the temperature of the groundwater is high, cold water may be used as the tracer.

5.8 Tracer container

The tracer container is the container that is used to prepare and store the tracer.

5.9 Fill equipment

The fill equipment is used to inject the tracer to the borehole.

Note: The tracer fill apparatus may be a vinyl hose, a portable miniature pump or other apparatus.

5.10 Stirring equipment

This equipment is used to stir the tracer inside the borehole to ensure that the water inside the borehole has a uniform electrical resistance or temperature.

Note: The stirring equipment used in the logging hole may be a compressor, an air chamber type air pump, a vinyl hose or other equipment.

5.11 Electrolyte

The electrolyte is used as the tracer solute in the case of logging by means of electrical resistance measurement.

5.12 Boiling equipment

The boiling equipment is used when hot water is used as the tracer.

6 Preparation of logging hole

The logging hole shall be prepared as follows.

a) Drill a borehole of the specified hole diameter [3] down to the specified depth [4].

Note [3]: The specified hole diameter shall be a diameter that enables the logging pipe to be installed and the hole interior to be flushed.

Note [4]: The specified depth shall be a depth that encompasses the logging section.

Note 1: Borehole shall be fresh-water drilling. However, if a mud conditioning agent is used, the borehole shall be flushed thoroughly.

b) Install the logging pipe into the borehole.

Note 2: When the borehole wall is free-standing, an open hole may be used.

c) Flush the logging hole. Pump up the water in the borehole using a bailer, an air lift, a pump or other apparatus and flush the interior of the hole repeatedly until the emulsified water becomes clear water. If mud fluid had to be used for drilling, the logging hole shall be flushed with particular care.

7 Logging method

7.1 Logging by means of electrical resistance measurement

Logging by means of electrical resistance measurement shall be performed as follows.

a) Using the water level meter, measure the water level in the borehole prior to the start of logging.

b) Install the logging cable.

Note 1: To ensure that the initial electrical resistance distribution is measured accurately, be careful not to disturb the water inside the borehole when the logging cable is installed.

c) Measure the electrical resistance distribution of the water in the borehole prior to the addition of the tracer.

Note 2: Confirm to make sure the initial measurement value is stable.

d) Dissolve the electrolyte in water to prepare the tracer.

Note 3: Adjust the concentration and quantity of tracer so the electrical resistance in the borehole becomes 1/5 - 1/10 of the value prior to the addition of the tracer.

e) Inject the tracer to the borehole and stir thoroughly so the electrical resistance of the entire quantity of water in the borehole is uniform.

Note 4: A compressor or an air chamber type air pump shall be used for stirring. Additionally, the method in which the vinyl hose used to add the tracer is moved up and down inside the borehole to stir the water while the tracer is being added may be used.

Note 5: Confirm that the electrical resistance of the water in the borehole is uniform.

f) Measure the water level inside the borehole when stirring has been completed.

g) Measure the electrical resistance of the water in the borehole over time at each measurement point.

Note 6: In general, the measurement times shall be immediately following agitation (0 minutes) and after 5 minutes, 10 minutes, 15 minutes, 30 minutes, 60 minutes, 90 minutes, 120 minutes, 150 minutes and 180 minutes. The upper limit for measurement time shall be approximately 180 minutes.

h) Measure the water level in the borehole when logging has been completed.

7.2 Logging by means of temperature measurement

Logging by means of temperature measurement shall be performed as follows.

a) Using the water level meter, measure the water level in the borehole prior to the start of logging.

b) Install the logging cable.

Note 1: Be very careful to avoid exposing the logging cable to direct sunlight and ensure that it does not become hot prior to insertion.

Note 2: To ensure that the initial temperature distribution is measured accurately, be careful not to disturb the water inside the borehole when the logging cable is installed.

c) Measure the temperature distribution of the water in the borehole prior to the injection of the hot water.

Note 3: Make sure the initial measurement value is stable.

d) Prepare the hot water.

Note 4: The temperature of the hot water shall be approximately 60 °C. Prepare a quantity of hot water sufficient to increase the temperature of the water in the borehole approximately 15 - 20 °C after the hot water is injected.

e) Add the hot water to the borehole and stir thoroughly so the temperature of the entire quantity of water in the borehole is uniform.

Note 5: A compressor or an air chamber type air pump shall be used for stirring. Additionally, the method in which the vinyl hose used to add the tracer is moved up and down inside the borehole to stir the water while the hot water is being added may be used.

Note 6: Confirm that the temperature of the water in the borehole is uniform.

f) Measure the water level inside the borehole when stirring has been completed.

g) Measure the temperature of the water in the borehole over time at each measurement point.

Note 7: In general, the measurement times shall be immediately following agitation (0 minutes) and after 1 minute, 2 minutes, 3 minutes, 4 minutes, 5 minutes, 6 minutes, 7 minutes, 10 minutes, 15 minutes, 20 minutes, 25 minutes, 30 minutes and 60 minutes. The upper limit for measurement time shall be approximately 60 minutes.

h) Measure the water level in the borehole when logging has been completed.

8 Analysis of test results

Annex A shows a method for analyzing the results.

9 Reporting

The following items shall be reported.

a) Logging borehole number, location and ground elevation, and pipe crown height

Note 1: The elevation should be determined and used as the ground height and the pipe crown height.

b) Water level in borehole

Note 2: The water level in the borehole before the start of measurement, after stirring has ended and after logging has been completed shall be reported.

c) Depth of logging section

d) Logging date / time and weather conditions

e) Logging method

f) Specifications of logging apparatus

Note 3: In the case of logging by means of electrical resistance measurement, the type of measurement that was conducted (electrical resistance or resistivity) shall be reported.

g) Interval between electrode pair or temperature sensors on logging cable

h) Electrical resistance or temperature of tracer and quantity of tracer added

i) Logging records

Note 4: The measurements for resistivity distribution or temperature distribution before and after the tracer or hot water was added shall be reported.

j) In the case of logging by means of electrical resistance measurement, the "Electrical resistance change - depth curve"

In the case of logging by means of temperature measurement, the "Temperature recovery rate - depth curve"

Note 5: If the boring log and other test results are available, these should be noted on the same drawing.

k) Results of groundwater flow layer determination

l) If the method used deviates in any way from this standard, give details of the method used.

m) Other reportable matters

Note 6: If necessary, the status of groundwater use in the surrounding area shall be reported.

Annex A
(Regulation)

Methods for analyzing results

A.1 Logging by means of electrical resistance measurement

The method used to analyze results from electrical resistance measurement shall be as follows.

a) The electrical resistance in the natural state and the changes over time in the electrical resistance after the tracer has been added shall be processed for each measurement point.

b) The electrical resistance immediately after the tracer has been added (0 minutes) and the difference in electrical resistance at each measurement time (electrical resistance change) shall be calculated for each measurement point.

c) Based on b), the "Electrical resistance change - depth curve" shown in Fig. B.1 in Annex B shall be prepared.

A.2 Logging by means of temperature measurement

The method used to analyze results from temperature measurement shall be as follows.

a) The temperature in the natural state and the changes over time in the temperature after the hot water has been injected shall be processed for each measurement point.

b) The temperature immediately after the hot water has been injected (0 minutes) and the difference in temperature at each measurement time t (temperature change) shall be calculated for each measurement point.

c) The temperature recovery rate at each measurement point shall be determined using the following equation.

$$t_r = \frac{\theta_d - \theta_t}{\theta_d - \theta_n} \times 100$$

where

t_r : Temperature recovery rate (%)
θ_d : Temperature immediately after addition of tracer (°C)
θ_n : Temperature in natural state (°C)
θ_t : Temperature at elapsed time t (°C)

d) The "Temperature recovery rate - depth curve" shown in Fig. B.2 in Annex B shall be prepared.

A.3 Determination of flow layer

Using the drawings that have been prepared in A.1 and A.2, the section in which significant changes are observed over time in electrical resistance or temperature shall be located in order to determine the groundwater flow layer.

Note 1: There is a rapid increase in electrical resistance or rapid temperature recovery in the section in Fig. B.1 and Fig. B.2 in Annex B in which the flow of groundwater is relatively rapid. For this reason, the shape of the curve in this part of the graph protrudes to the right. This shall be determined as the groundwater flow layer.

Note 2: If there are other test or logging results or the like, these shall be taken into consideration as well when determining the flow layer.

Annex B
(Reference)

Example of analyzing the results

B.1 Example of electrical resistance change - depth curve

Figure B.1 shows an example of electrical resistance change - depth curve.

B.2 Example of temperature recovery rate - depth curve

Figure B.2 shows an example of temperature recovery rate - depth curve.

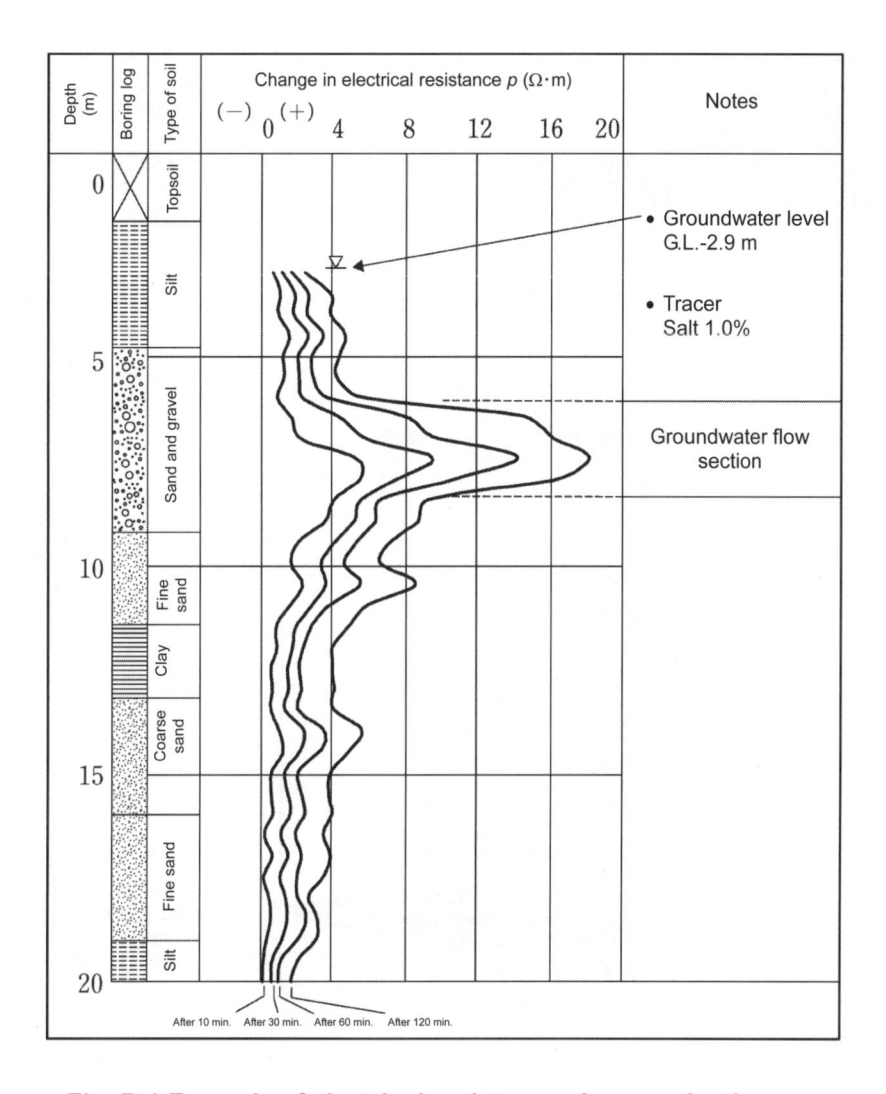

Fig. B.1 Example of electrical resistance change - depth curve

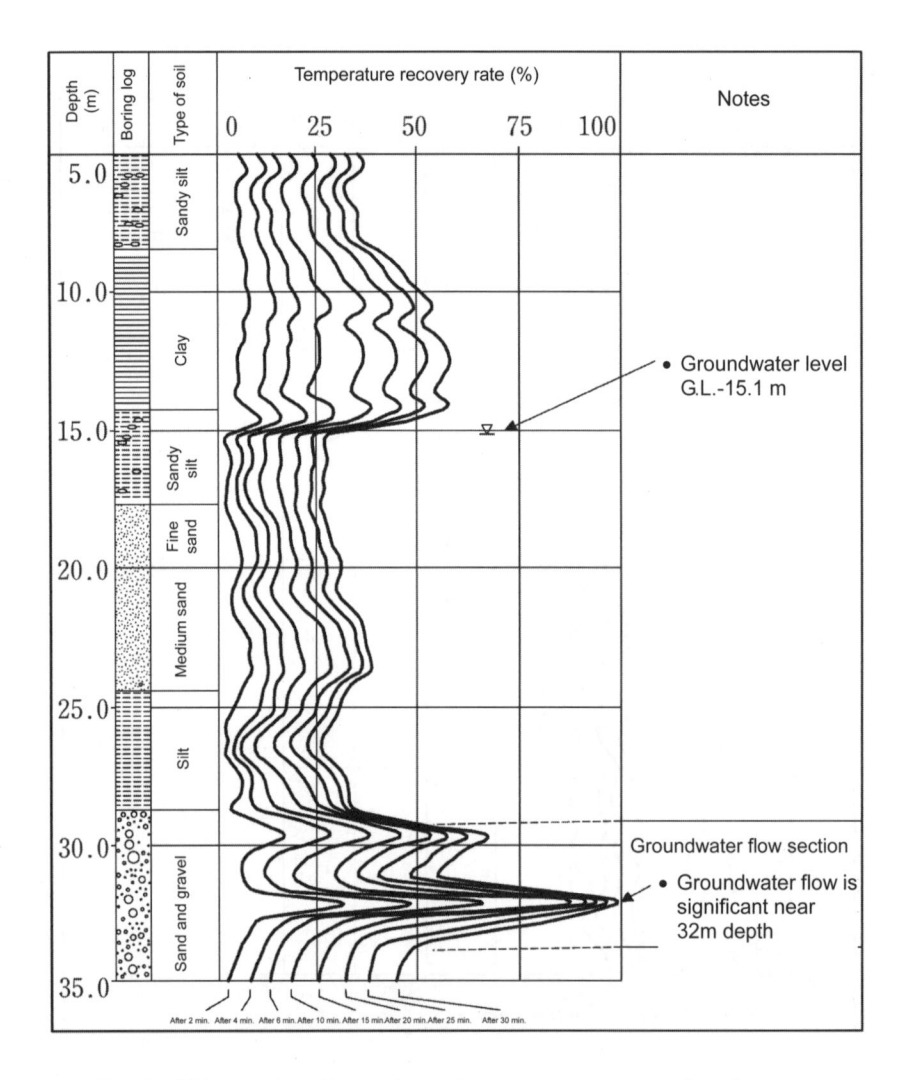

Fig. B.2 Example of temperature recovery rate - depth curve

Japanese Geotechnical Society Standard (JGS 1318-2015)
Method of detection of direction and velocity of groundwater flow in single borehole

1 Scope

This standard specifies a method for detecting the direction and velocity of groundwater flow in a single borehole.

This method applies to saturated ground beneath the groundwater level, and detects the direction and velocity indirectly using borehole water. This method does not apply when the opening rate of the screen installed in the measurement section is less than 10%.

2 Normative references

The following standard shall constitute a part of this standard by virtue of being referenced herein. The latest version of this standard shall apply (including supplements).

JGS 1311   Method for measuring groundwater level in borehole

3 Terms and definitions

The main terminology and definitions used in this standard shall be as follows:

3.1 Method for detection of direction and velocity of groundwater flow

Method for detecting the direction and velocity of groundwater flow by introducing various kinds of tracer into a measurement section, and measuring the variation with time of the electrical resistance or water temperature of the borehole water, or measuring the positions of the tracers moving in the water flow within the borehole.

3.2 Tracer

Water having different electrical resistance from that of the borehole water, a heat source that is added to the borehole water directly, or solid fine particles that move in the flow of the borehole water.

3.3 Electric Potential difference measuring method

A method of detecting electric potential difference by potential difference distribution using a solution tracer.

3.4 Temperature measurement method

A method of detecting temperature differences by adding heat using a temperature tracer.

3.5 Particle tracking method

A method of detecting particle position using a solid fine particle tracer.

3.6 Opening rate

The percentage of the borehole water inflow part on the screen, such as a pipe in which slits or circular holes are formed.

3.7 Measurement velocity

Velocity of groundwater flow through the borehole as measured by a flow direction and velocity measuring device.

Note: Sometimes the measured value differs from the velocity of groundwater flow through the ground, so appropriate conversion may be necessary.

3.8 Darcy's velocity

The apparent velocity that is averaged over the cross-sectional area of the total of the voids and soil particles.

4 Equipment

The measuring equipment using this standard shall be as follows:

4.1 Drilling equipment

A boring machine capable of drilling to a specified depth at a specified diameter.

4.2 Screen

The protective pipe with slits or circle holes for measuring the flow of the groundwater.

Note: The screen may not be used in cases where a protective pipe has been installed to above the measurement section, or in measurement holes in ground where the borehole walls are stable.

4.3 Filter material

Material filled between the borehole wall and the screen

Note: In general, silica sand or small gravel is used. In addition, filter material is not used in open holes.

4.4 Net

When there is a possibility of inflow of filter material or soil particles into the screen because the slits or circular holes are large, the net has a function of preventing this inflow without preventing the flow of water in the screen.

4.5 Flushing equipment

Equipment to clean the inside of the borehole.

Note: Cleaning equipment includes a bailer, air lift device, a pump, etc.

4.6 Packer

When a component of vertical flow occurs in a borehole, the packer seals between the measuring instrument and the borehole or screen by expanding rubber or similar attached to the measuring instrument. In addition, it is used to stabilize the measuring instrument.

4.7 Measuring instrument

An instrument that measures flow direction and velocity by the potential difference method, the temperature method, or the particle tracking method.

4.8 Direction sensor

4.8.1 Built-in direction meter

A sensor that detects direction.

Note: Generally, this sensor is attached to a measuring instrument.

4.8.2 Constant azimuth rod

A rod with connections having an insertion structure that enables alignment of the direction of the measuring instrument.

4.9 Tracer

4.9.1 Electric potential difference method

Water having a different electric potential from that of the borehole water.

Note: In general, distilled water is to be used as a tracer, but a salt solution of table salt, etc., may be used.

4.9.2 Temperature method

Water having a different temperature from that of the borehole water.

Note: In general, heat source with a higher temperature than that of the groundwater shall be used.

4.9.3 Particle tracking method

Solid fine particles that move with the borehole water flow.

5 Preparation of measurement hole

The procedures are as follows for measurement in an open hole and within a screen, respectively:

5.1 Measurement in an open hole

Measurement borehole preparations shall be as follows.

a) Excavate and shield water to the upper part of measurement section in accordance with JGS 1311, and install casing tube with a specified borehole diameter [1] to the top of the measurement section.

Note [1]: Borehole diameter sufficient to enable installation of a measuring probe and flushing inside the borehole.

Note 1: When the borehole wall is free-standing, an open hole may be used.

Note 2: Fresh water drilling is desirable, but when the use of mud fluid is necessary for borehole wall protection, replace with fresh water after protective tube installation.

b) Drill to the specified depth of the measurement section [2] with the specific borehole diameter.

Note [2]: The specified depth is the depth that covers the measurement section fully.

c) Flush the measurement hole. Flush the borehole until water with suspended solids turns into clear water, by raising the borehole water with a bailer, a pump, air lift [3] etc. Flush it carefully when it has been necessary to use mud fluid for drilling. Further, if there is sufficiently low prospect for recovery of water level in a flushed hole of low hydraulic conductivity, inject fresh water to the borehole.

Note [3]: An air lift is apparatus that withdrawal borehole water with rising air bubbles by discharging compressed air around the tip of a lift pipe that is deeply installed in the borehole.

Note 3: Ensure that water lifting does not affect the stability of the borehole wall, by taking care over the water lifting rate.

Note 4: When using air lift, discharge air at a depth which does not affect the measurement section.

Note 5: Take care that the water level in the borehole does not become higher than that of the surrounding groundwater level, at the time of injecting fresh water.

5.2 Measurement within a screen

Measurement hole preparations shall be as follows.

a) Drill to the specified depth [2] with the specified borehole diameter [1].

Note [1]: Specified bore size is the bore size of a screen pipe that enables a measuring instrument to be installed and the inside of the borehole to be flushed, or a borehole diameter in which the filter thickness is taken into consideration.

Note 2): The specified depth is the depth that covers the measurement section fully.

Note 1: Fresh water drilling is desirable, but when the use of mud fluid is necessary for borehole wall protection, replace with fresh water after protective tube installation.

b) Install screen pipe [3)] vertically in the drilled hole.

Note 3): The opening rate and the shape of the screen should be set so as not to prevent inflow of groundwater and so that the inflow of fine grained soils from the filter material and ground is as low as possible.

c) Fill filter material[4)] into the gap between borehole wall and pipe at the depth of the screen.

Note 4): The filter material should have sufficient permeability and reduce the inflow of fine grained soils from the ground as much as possible.

d) Shield the water with seal material [5)] in the gap between the screen and the borehole wall so that water flow in the vertical direction which could affect the measurement does not occur.

Note 5): Seal materials include bentonite and chemical expansive admixtures, and must be selected in accordance with measurement period, the water quality, the geological conditions, etc.

Note 2: The length of the screens should be set so that it is in the measurement section only, so that water flow in the vertical direction which could affect the measurement does not occur.

e) Flush the measurement hole. Flush the borehole until water with suspended solids turns into clear water, by raising the borehole water with a bailer, a pump, air lift etc. Flush it carefully when it has been necessary to use mud fluid for drilling. Further, if there is sufficiently low prospect for recovery of water level in a flushed hole of low hydraulic conductivity, inject fresh water to the borehole.

Note 3: When using air lift, discharge air at a depth which does not affect the measurement section.

Note 4: Take care that the water level in the borehole does not become higher than that of the surrounding groundwater level, at the time of injecting fresh water.

6 Measurement method

6.1 Electric potential difference method

When the electric potential difference method is used, the measurement method shall be as follows.

a) Measure the water level or water pressure in the borehole before setting measuring instrument.

b) Install measuring instrument to the specified depth and confirm the relationship between measurement equipment and its orientation.

Note 1: To set the orientation, it is desirable to use direction sensors, etc.

c) Inflate packers to shield the measurement section from its upper and lower parts and shut out water flow.

Note 2: The packer inflation pressure should be set taking into consideration the water pressure in the borehole, and care shall be taken not to damage borehole wall.

Note 3: It is desirable to set the packers at the depth of protective tubes that are set at the top and the bottom of the screen.

Note 4: If it is not necessary to partition within the measurement section, it is not necessary to shield with a packer.

d) Confirm that the water level or water pressure inside the borehole is stable before starting measurement.

Note 5: Confirm that the effect on the water flow due to the installation of the measuring instrument has disappeared by continuous measurement of the change in the water pressure or electric potential difference.

e) Start measuring by injecting a tracer solution (distilled water, etc.), whose electrical resistance is different from that of borehole water, from the central part of the measuring instrument.

Note 6: Take care that tracer injection does not affect groundwater flow.

f) Measure changes of the electric potential value over time with electrodes set on the measuring instrument.

Note 7: Make the measuring interval (around 1 second, but depends on the capability of the measuring instrument) after tracer injection as short as possible, and the measurement interval may be lengthened according with the change in the amount of the electric potential difference.

g) Finish measuring after confirming a change in the electric potential value (detecting a peak).

6.2 Temperature method

There are 2 types of temperature method, the heating type and the heat flow type, and these methods are as follows.

6.2.1 Heating type measurement

The heating type of temperature method is as follows.

a) Preparations for this method shall be in accordance with 6.1 a) - b).

b) Before starting the measurement, confirm the values of the thermometer sensor and the direction sensor on the measuring instrument.

c) Measure the temperature of the groundwater in the natural condition.

Note 1: When measuring the natural condition, make sure there are no changes in the temperature.

d) Apply voltage to the heater and measure the change in temperature with time via the thermometer in the measuring instrument.

Note 2: When measuring after applying the voltage, the measurement interval may be extended in accordance with the state of the temperature changes.

e) After measuring for a certain period of time and the trend of groundwater flow has been determined, stop applying voltage to the heater.

f) Measure the rate of the drop in temperature over the next several minutes then finish.

6.2.2 Heat flow type measurement

The heat flow type of temperature method is as follows.

a) Preparations for this method shall be in accordance with 6.1 a) - c).

b) Confirm that the temperature distribution around the measuring instrument is stable before measurement.

Note: Confirm that the effect of installation of the measuring instrument on the groundwater flow has been eliminated by continuously measuring the temperature change of the groundwater.

c) Heat the heat source installed in the center of measuring instrument, then start the measurement.

d) Continuously measure the temperature change of multiple temperature sensors around the measuring instrument, and the direction and velocity of groundwater flow.

e) When the distribution of the direction and velocity of the groundwater flow has stabilized, stop the measurement.

6.3 Particle tracking method

There are 2 types of particle tracking method, the optical type and the ultrasonic type, and these methods are as follows.

6.3.1 Optical type measurement

a) Preparations for this method shall be in accordance with 6.1 a) - d).

b) Solid fine particle tracer shall be injected in the measurement section without affecting the groundwater flow, then measurement shall be started.

Note 1: Tracer shall be injected in the center of the measurement section.

c) The fine particle tracer moving in the measurement section shall be measured by image capture.

Note 2: The measurement time interval shall depend on the flow velocity, e.g. 20-40 seconds as a standard measurement time in the case of 1×10^{-4} (m/s) flow velocity within the hole.

Note 3: Repeated measurement is recommended for improving the reliability of the measurement results.

Note 4: It shall be confirmed that there are no large changes in the water pressure and temperature during the measurement.

d) The tracer position shall be measured as coordinates in space by acquiring image data from variable directions.

e) The measurement shall be finished when the tracer travel length and direction of travel path have been confirmed.

Note 5: The measurement can be finished in 170 minutes from start of measurement when the tracer has traveled 1 cm length at 1×10^{-6} m/s flow velocity.

6.3.2 Ultrasonic type measurement

a) Preparations for this method shall be in accordance with 6.1 a) - d).

b) Start measuring by injecting the solid fine particle tracer into the measurement section.

Note 1: Inject a sufficient quantity of tracer mixture water so that the distribution of tracer in the measurement section is uniform at the time of injecting the tracer.

c) Measure the behavior of the tracer moving in the borehole water as coordinates by measuring the reflected waves from the tracer with an ultrasonic sensor.

Note 2: The measurement interval depends on the instrument specification and the flow velocity in the borehole water.

Note 3: It is desirable that multiple coordinate values are acquired for at least a single particle.

Note 4: During the test, confirm that the water level, water pressure and water temperature inside borehole have been stable.

d) Finish measuring after confirming the travel distance and travel path direction of tracer that could be measured from multiple coordinate values.

Note 5: From the relationship between travel distance that can be measured and time, if for example the flow velocity within the hole is 10^{-5} m/s, it takes about 17 minutes for a travel distance of 1 cm, and if the flow velocity within the hole is 10^{-10} m/s it takes about 4.7 days to travel 40 μm.

7 Analysis of test results

The method of analyzing of the results is shown in Annex A.

8 Reporting

The following items shall be reported

a) Name and location of measurement borehole, and ground elevation

Note 1: The ground level should be obtained from the elevation.

b) Water level or pressure in borehole

Note 2: These should be reported as the water level or water pressure in the borehole after the end of measurement and before start of measurement.

c) Measurement depth

d) Measurement date and weather conditions

e) Measurement method

f) Specifications of the measuring instrument

g) Measurement records

Note 3: These should be reported as the change of tracer with elapsed time.

h) Measured direction and velocity of groundwater flow

i) Other reportable matters

Note 4: If necessary the situation of groundwater use in the surrounding area should be reported.

Annex A
(Requirement)

Method of analyzing the results

A.1 Electric potential difference method

The method of analyzing the results for the electric potential difference method shall be as follows.

a) Analyze the change with time of each electrode potential value after injecting tracer, as shown in Figure B.1 of Annex B.

b) Judge the direction of water flow from the change of each electrode with time, by choosing the electrode that has a biggest change in the electric potential at its peak. Also, calculate the water flow velocity from the time to reach the peak of the electrode chosen for judging the direction, from the following equation.

$$V_0 = \frac{L}{2t}$$

where

V_0: Measured water flow velocity
L: Distance between electrodes
t: Elapsed time to the peak

c) The Darcy flow velocity is calculated as follows.

$$V_d = V_0 \times \beta$$

where

V_d: Darcy's velocity
V_0: Measured water flow velocity
β: Correction coefficient (obtain the correction coefficient β from Table A.1)

A.2 Temperature method

The temperature method includes the heating type and the heat flow type, and analysis of the results is as follows.

A.2.1 Heating type measurement

Analysis of the results in case of the heating type shall be as follows.

a) Analyze the changes over time for each thermometer in the measuring instrument as shown in Figure B.2.1 of Annex B after applying the prescribed voltage to the heater of the internal measuring instrument sensor.

b) Measure the difference of the temperature of the temperature sensors around the heater before the application of voltage and the temperature while applying voltage, for various flow velocities and screens.

c) Calculate the flow velocity by substituting the difference of the average temperature before applying voltage and the average temperature while applying voltage into the equation obtained in b). Also, from the difference of the measured temperatures at an arbitrary time, produce a temperature distribution diagram on an isothermal chart, and determine the flow direction.

A.2.2 Heat flow type measurement

Analysis of the results for the heat flow type shall be as follows.

a) Measure the temperature distribution at each point after heating the heat source, and analyze the groundwater flow direction and velocity as shown in Fig B.2.2 of Annex B.

b) Darcy's flow velocity is obtained from the following equation.

In the flow velocity region $2 \times 10^{-6} \leq V_d \leq 10^{-4}$ m/s

$$V_d = [T_{dev}/61.33]^{1.16}$$

In the flow velocity region $V_d \geq 10^{-4}$ m/s

$$V_d = [\Delta T/0.053]^{-1.19}$$

where

V_d: Darcy's flow velocity
T_{dev}: Standard deviation of the average temperature rise
ΔT: Average temperature rise

A.3 Particle tracking method

The particle tracking methods include the optical type and the ultrasonic type, and analysis of the results is carried out as follows.

a) Analyze the tracer movement after injecting the tracer as coordinate values, as shown in Figure B.3.1 and Figure B.3.2 of Annex B.

b) Judge the direction of groundwater flow from the paths of the tracer. Also calculate the velocity of groundwater flow from the following equation from the movement distance of the tracer per unit time, obtained from the coordinate values used to judge the direction of groundwater flow.

$$V_0 = \frac{L}{t}$$

where

V_0: Velocity of borehole water flow
L: Movement distance of tracer
(straight line distance from base-point coordinates x,y,z to x',y',z'

$$L = \sqrt{(x-x')^2 + (y-y')^2 + (z-z')^2}\ )$$

t: Time required for movement of tracer

Table A.1 Relation between hydraulic conductivity and correction coefficient

Material	D_{10} (mm)	D_{20} (mm)	D_{50} (mm)	Hydraulic conductivity k (m/s)	Correction coefficient β
ø0.1 mm Beads	0.10	0.10	0.10	4.9E-05	0.159
Toyoura sand	0.17	0.18	0.20	1.8E-04	0.204
Coarse sand I	0.50	0.63	0.90	1.5E-03	0.268
Coarse sand II	0.90	1.20	1.35	5.9E-03	0.395

Annex B
(Reference)

Example of analysis of results

B.1 Example of the electric potential difference method

Figure B.1 shows an example of the electric potential difference method.

B.2 Example of the temperature method

Figures B.2.1 and B2.2 show examples of the heating type and heat flow type, respectively.

B.3 Example of the particle tracking method

Figures B.3.1 and B3.2 show examples of the optical type and ultrasonic type, respectively.

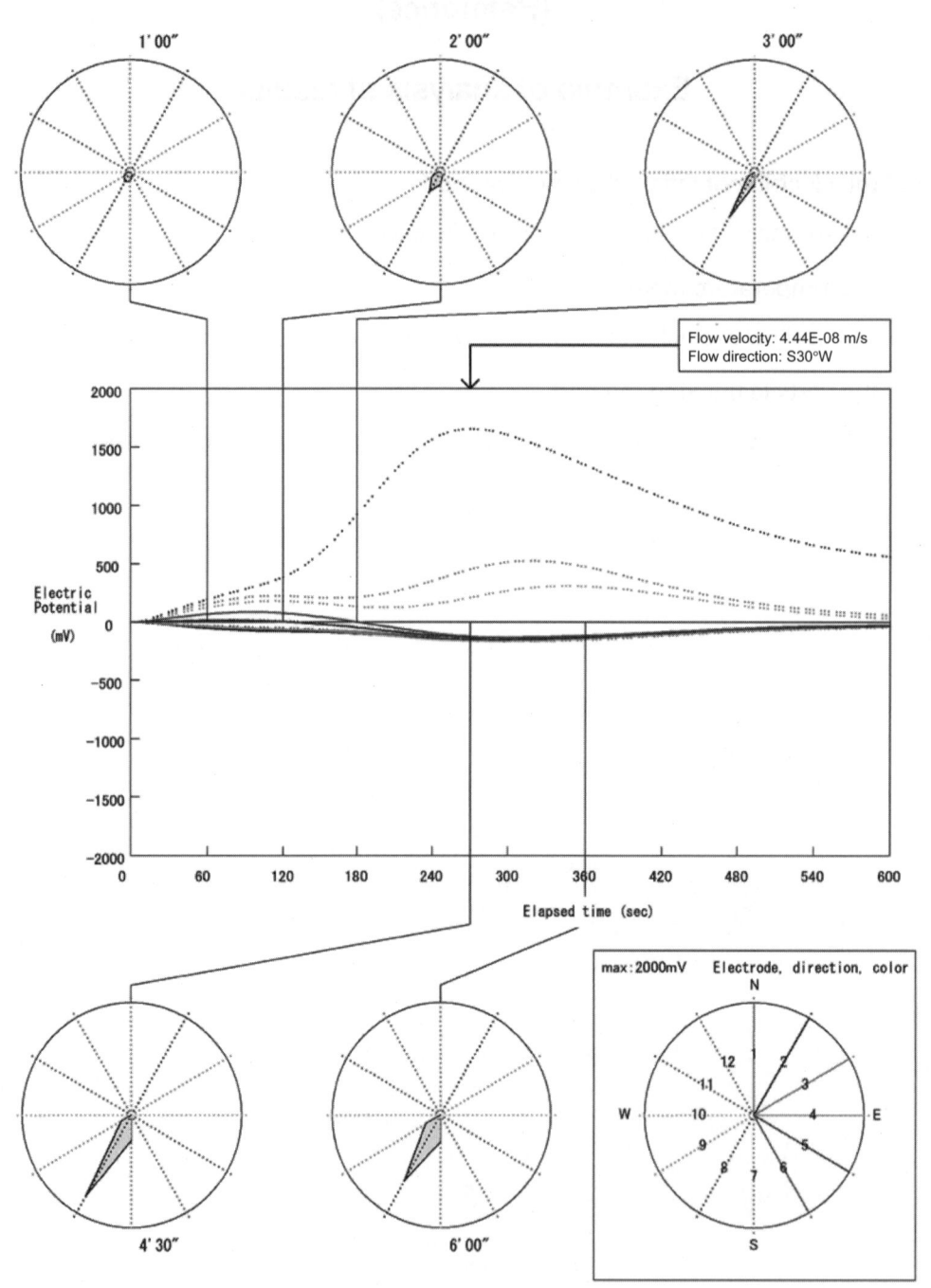

Figure of the result of 12 components groundwater flow velocity and direction

Flow velocity: 4.44E-08 m/s
Flow direction: S30°W

Fig. B.1 Example of the electric potential difference method

Example of the flow velocity.

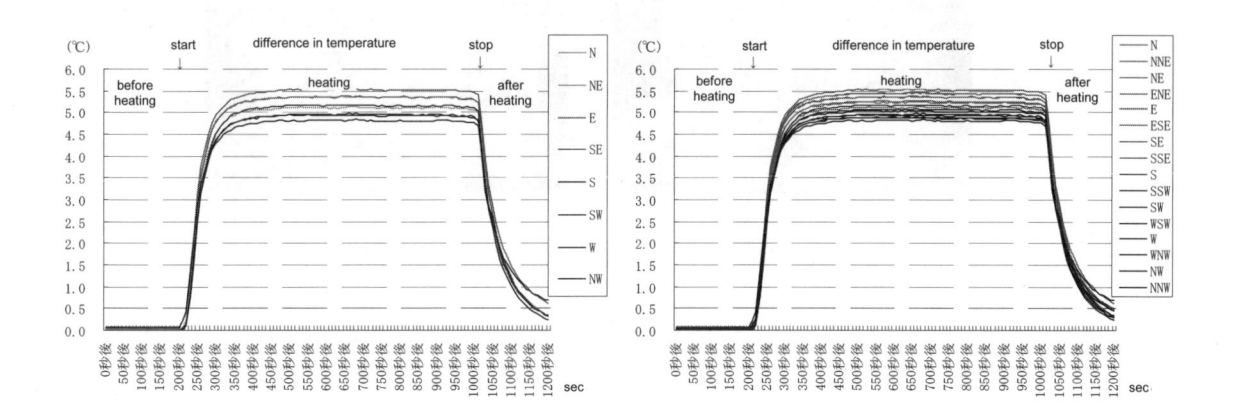

Example of the flow direction.

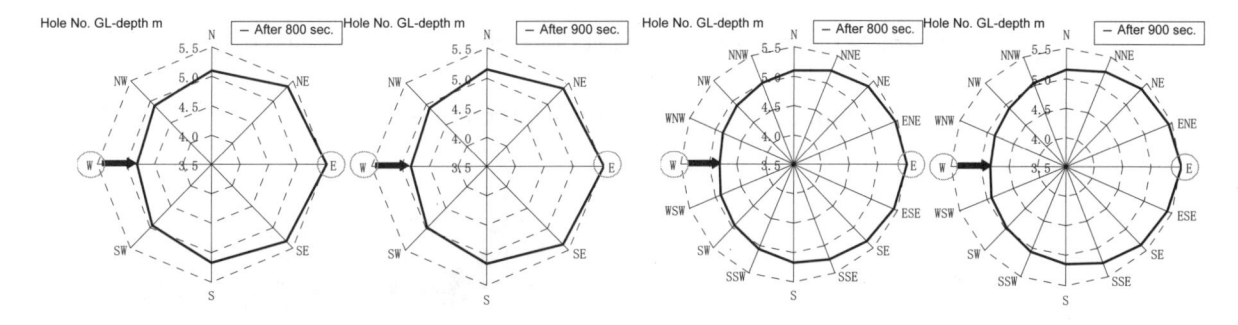

Fig. B.2.1 Example of heating type.

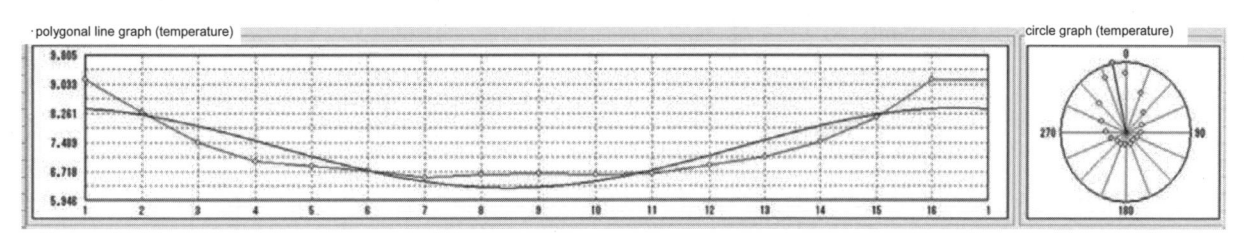

Fig. B.2.2 Example of heat flow type.

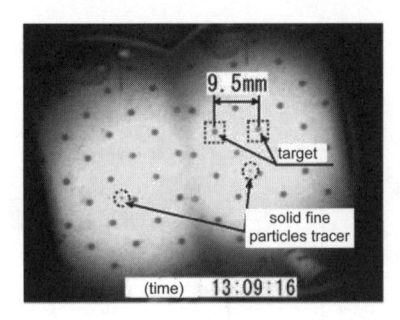

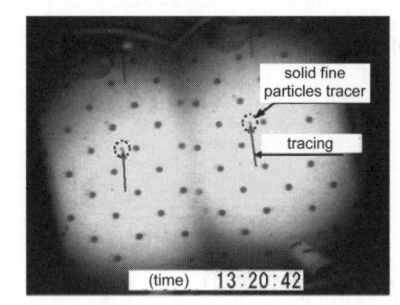

(a) Example of image acquisition varying the photography angle in the right and left images.

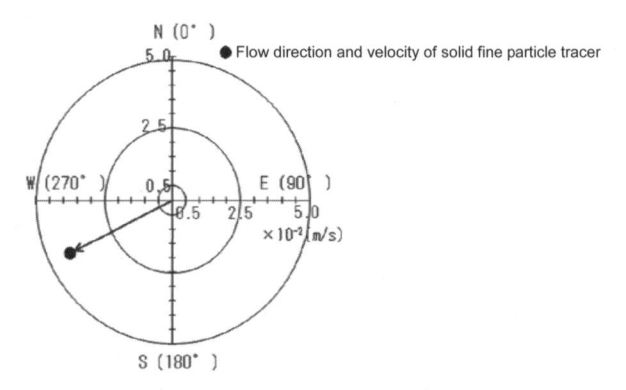

(b) Example of measurement result (flow direction and velocity).

Fig. B.3.1 Example of the optical type.

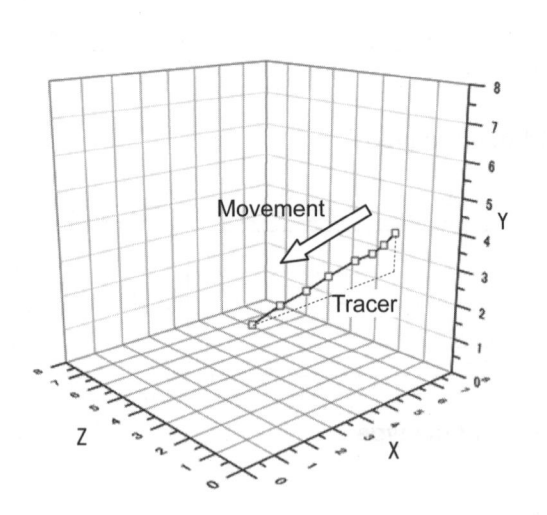

(a) Image of a coordinate values of the movement trajectory of tracer

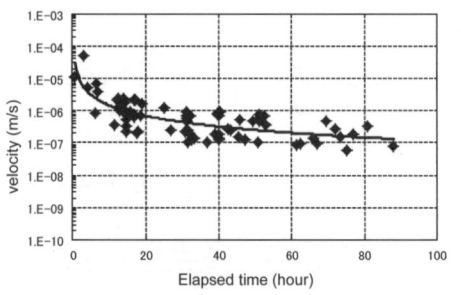

(b) Example of measurement result (flow velocity)

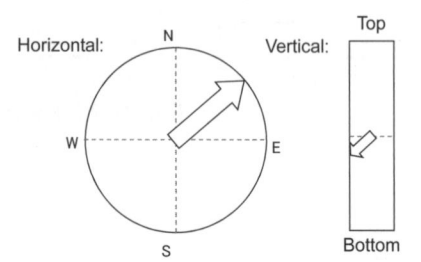

(c) Example of measurement result (flow direction)

Fig. B.3.2 Example of the ultrasonic type

JGS 1321-2012

Japanese Geotechnical Society Standard (1321-2012)
Method for determination of hydraulic properties of rock mass using instantaneous head recovery technique in single borehole

1 Scope

This standard specifies methods for determining the equilibrium water level and hydraulic conductivity of a rock mass using a single borehole. The test applies to saturated rock mass beneath the level of groundwater.

Note 1: This test is suitable for rock mass whose hydraulic conductivity is approximately 10^{-4} m/s or less.

Note 2: In this test, the coefficient of hydraulic conductivity shall be determined by assuming that the rock mass is a homogenous porous medium.

Note 3: This test is suitable when approximately 90% of the initial water level differential is recovered during the duration of the test, and when at least 10 points of observation data can be obtained during the test.

2 Normative references

None

3 Terms and definitions

The main terms and definitions used in this standard are as follows:

3.1 Hydraulic test of rock mass using groundwater level recovery method in single borehole

This is a transient hydraulic conductivity test method in which a water level measurement pipe with a packer and a trip valve attached on the end is installed into the borehole and the packer is inflated to demarcate an arbitrary test section, after which the trip valve is opened to restore the water level inside the water level measurement pipe and the coefficient of hydraulic conductivity is determined from the relationship between the water level recovery in the pipe and the time, and the equilibrium water level is determined from the observed value when water level recovery has stopped.

4 Equipment

Note: Fig. B.1 in Annex B shows a simplified example of the test equipment.

4.1 Drilling equipment

The drilling equipment shall comprise a boring machine that can drill rock mass and clean the interior of the borehole.

4.2 Water level measurement pipe

The water level measurement pipe shall be a watertight pipe that can measure the water level in the test section.

Note 1: The connecting sections of the water level measurement pipe shall be provided with sealing tape or other means to prevent leakage.

Note 2: The inner diameter of the water level measurement pipe shall be uniform throughout the water level recovery section.

4.3 Trip valve

The trip valve shall be attached to the water level measurement pipe and shall be equipped with a feature that enables it to be opened instantaneously when the test is started.

Note 1: Normally the inner diameter of the trip valve shall be approximately 20 - 30 mm, but the value shall be determined with consideration for the occurrence of turbulent flow, the resistance inside the pipe and other factors.

Note 2: When a water level measurement cable equipped with electrodes at fixed intervals is used to measure the water level, a cable with a weight attached to the end shall be used to open the trip valve.

4.4 Packer

The packer shall adhere tightly to the borehole wall and shall not produce water leakage during the test.

Note 1: The types of packers comprise a single packer that is placed above the test section and a double packer that is placed both above and below the test section. The type shall be selected based on the status of the borehole wall and the test procedure.

Note 2: The packer should be approximately 1 m or more in length.

4.5 Pressurizing apparatus

The pressurizing apparatus shall be able to inflate the packer using gas pressure or hydraulic pressure.

4.6 Water level meter

The water level meter shall be able to measure and record the water level in the water level measurement pipe over time.

Note 1: A water pressure gauge may be used to measure the water level. In such cases, the gauge should be able to measure the water level in 10 mm increments and should be able to automatically record the time in seconds.

Note 2: The calibration coefficient and the indicator value in no-load status shall be checked in advance, and the water pressure gauge shall be calibrated prior to use.

5 Preparation of test hole

Test hole preparations shall be as follows.

a) Select a location for the borehole in an area that will not be affected by existing horizontal shafts, boreholes and so on.

b) Use the drilling equipment to drill the borehole.

Note 1: The test hole should be drilled using fresh water. However, if slurry is used, the borehole shall be flushed thoroughly

Note 2: In order to install the packer, trip valve, water level measurement pipe and so on, and in order to maintain the shape of the test section, the finished borehole shall have stable borehole walls that will not collapse.

c) Flush the borehole until there is no slime in the return water.

6 Test method

6.1 Test preparations

The following test preparations shall be made.

a) Determine the depth of the test section in accordance with the geological conditions of the rock mass being tested.

Note 1: The L/D ratio between the length L of the test section and the diameter D of the borehole in the test section shall be 4 or greater. If $L/D \geq 4$ cannot be secured, the test results shall be organized using a different equation from the one shown in A.2 in Annex A.

b) Measure the water level inside the borehole.

Note 2: The measured water level inside the borehole shall be used to determine the location of the trip valve.

c) Attach the packer to the end of the water level measurement pipe. Attach the trip valve to the top of the packer.

d) While connecting the water level measurement pipe to which the packer and trip valve have been attached, install the pipe down to the depth determined in a). Normally the trip valve shall be closed during this process.

e) Using the pressurizing apparatus, inflate the packer so it adheres to the borehole wall.

Note 3: If gas pressure is used to inflate the packer, apply sufficient gas pressure to prevent water leakage during the test, taking into consideration the depth of the test section, the equilibrium water level in the test section (if this can be measured in advance), the water level in the borehole, the effective expansion pressure of the packer (the pressure used to maintain the water shielding effectiveness of the packer) and the water level differential applied during the test.

Note 4: An effective way of confirming the water shielding effectiveness of the packer is to measure the water level inside the borehole above the packer (the water level in the gap between the water level measurement pipe and the borehole wall) before and after the test. If a significant increase in water level is observed before or after the test, this is thought to indicate that the water shielding effectiveness of the packer is inadequate. Change the position of the packer and conduct the test again.

Note 5: Measure the pressure of the gas or water used to pressurize the packer before and after the test. If there is a significant change in pressure before or after the test, there may be leakage from the packer or the piping system. In such cases, raise the packer and conduct inspection or replacement.

6.2 Test method

The test method shall be as follows.

Note 1: An overview of the test method is shown in Fig. B.2 in Annex B.

a) After the water pressure in the test section has stabilized, open the trip valve and start measurement.

Note 2: Install a water pressure gauge beneath the trip valve to confirm that the water pressure in the test section has stabilized. Alternately, a configuration that permits the trip valve to be opened and closed from ground level is needed.

Note 3: If a test apparatus with the configuration indicated in Note 2 is used, measure the equilibrium water level in the test section before conducting the hydraulic conductivity test. In such cases, a pump, air lift or other means shall be used to set an appropriate water level differential at the start of the test, using the measurement of equilibrium water level as a reference. Care shall be taken during this process to ensure that the water level differential at the start of the test is not excessively high, leading to a turbulent state.

Note 4: If a test apparatus with the configuration indicated in Note 2 cannot be used, wait a sufficiently long period of time for the water pressure in the test section to stabilize and then open the trip valve and start the test.

b) After opening the trip valve, measure the values over time for the water level h (m) inside the water level measurement pipe and the time t (s).

Note 5: The test start time $t = 0_s$ shall be the point at which the trip valve has been opened, or the point at which the decrease in the water level in the water level measurement pipe is greatest as determined by water level observations beginning prior to the start of the test. The water level at this time shall be test start water level h_p (m).

Note 6: If a water pressure gauge is used, the water pressure shall be measured at the specified times and the values shall be converted into water level values. The depth at which the water pressure gauge has been placed shall be recorded as well.

c) Continue the test until the water level recovery is less than 10 mm per hour. Record the final water level as equilibrium water level h_0 (m).

7 Analysis of test results

The results shall be analyzed in terms of the relationship between the water level h (m) to which the water level in the water level measurement pipe has recovered and the measurement time t (s). An $h-t$ curve like the one shown in Fig. B.3 in Annex B shall be prepared. A method for analyzing the results is shown in Annex A.

8 Reporting

The following items shall be reported.

a) Number and location of borehole, and ground elevation

Note 1: The elevation should be determined as the ground height.

b) Depth, borehole diameter and length of test section

c) Geological conditions of test section

d) Test date and time, weather conditions and water level in borehole

e) Configuration of test apparatus

f) Equilibrium water level in the test section

g) Water level measurement records

h) Method of processing the data and the corresponding water level recovery curves

Note 3: Show as a curve of $\log_{10} s\text{-}t$ or $s/s_p\text{-}\log_{10} t$.

i) Hydraulic conductivity

Note 4: If the type curve matching method is used for analyzing the test results, report the specific storage Ss (1/m).

j) If a method that partially differs from this standard was used, details of the points of difference

k) Other reportable matters

Annex A
(Regulation)
Methods for analyzing results

A.1 Equilibrium water level

The water level h (m) to which the level in the water level measurement pipe has recovered and the measurement time t (s) shall be plotted as shown in Fig. B.3 in Annex B, and the equilibrium water level h_0 (m) at which water level recovery has almost completely stopped shall be determined from the asymptotic line in the $h-t$ curve.

A.2 Straight line method

The straight line method is used when it has been determined that the test results have not been affected by the retention capability of the rock mass. This method is used when a straight line section is observed in the $\log_{10} s - t$ curve in Fig. B.4 in Annex B.

The straight line method can be used to determine only the hydraulic conductivity.

The following method shall be used to analyze the results of the straight line method.

a) On a semi-log graph, plot the difference in water level $s\,(= h_0 - h)$ (m) between the equilibrium water level h_0 and the water level h (m) in the water level measurement pipe on the logarithmic scale (vertical axis) and the time t (s) on the arithmetic scale (horizontal axis), in order to create a $\log_{10} s - t$ curve as shown in Fig. B.4 in Annex B. Check to see if a straight line gradient can be observed in the plotted graph.

b) Determine the gradient a (1/s) of the straight line that has been observed. The gradient a shall be determined from the coordinates of two arbitrary points (t_1, $\log_{10} s_1$) and (t_2, $\log_{10} s_2$) on the straight line, using the following equation.

$$a = \frac{\log_{10}(s_1/s_2)}{t_2 - t_1}$$

c) Calculate the hydraulic conductivity k (m/s) using the following equation.

$$k = \frac{(2.3 d_e)^2}{8L} \log_{10}\left(\frac{2L}{D}\right) a$$

for

$$\frac{L}{D} \geq 4$$

where

d_e: Diameter of circle with an area equivalent to the effective cross-sectional area determined by subtracting the cross-sectional area c (m^2) of the water level measurement cable from the cross-sectional area of the interior of the water level measurement pipe

$$\left(= \sqrt{d^2 - \frac{4c}{\pi}}\right) \text{ (m)}$$

However, $d_e = d$ if a water pressure gauge is used and a cable is not installed into the water level measurement pipe

d : Inner diameter of water level measurement pipe (m)
D : Diameter of borehole in test section (m)
L : Length of test section (m)

A.3 Type curve matching method

The type curve matching method is used when it is determined that the test results have been affected by the retention capacity of the rock mass and there is no clear straight line section in the $\log_{10} s - t$ curve.

The type curve matching method can be used to determine the hydraulic conductivity and the specific storage.

The following method shall be used to analyze the results obtained through the use of the type curve matching method.

Note: The specific storage that is determined may vary considerably depending on the method used to process errors and differences in the method used for type curve matching.

a) From the equilibrium water level h_0 (m), the water level h (m) measured during the test and the water level h_P (m) at the start of the test, determine the difference in water level $s\, (= h_0 - h)$ (m) during the test and the difference in water level $s_P\, (= h_0 - h_P)$ (m) at the start of the test. Also determine the water level differential ratio s/s_P.

b) Plot the measurements on a semi-log graph, with the water level differential ratio s/s_P as the arithmetic scale (vertical axis) and the elapsed time t (s) since the start of the test as the logarithmic scale (horizontal axis).

c) On a different semi-logarithmic graph with the same scale as the graph created in b), create a group of type curves showing the relationship between the water level differential ratio s/s_P for each storage coefficient ratio α and dimensionless time β, as shown in Fig. B.5 in Annex B.

d) Overlay the two graphs created in b) and c) as shown in Fig. B.6 in Annex B. Move one graph in parallel in the direction of the time axis (horizontal axis) and select the type curve that most closely matches the measured values. Read the value for α (α_m) that corresponds to this standard curve and the time axis coordinates t_m and β_m for both graphs that correspond to an arbitrary matching point.

e) Determine the hydraulic conductivity k (m/s) and the specific storage S_S (1/m), using the following equations.

$$k = \frac{d_e^2 \beta_m}{4L t_m}$$

$$S_S = \frac{d_e^2}{LD^2} \alpha_m$$

Annex B
(Reference)

Example of test apparatus and analysis of test results

B.1 Example of test apparatus

Fig. B.1 shows a simplified example of the test apparatus.

B.2 Example of test method

Fig. B.2 shows an example of the test method.

B.3 Relationship between water level and water level measurement time

Fig. B.3 shows an example of the relationship between the water level h and the water level measurement time t.

B.4 Log curve

Fig. B.4 shows an example of the $\log_{10} s - t$ curve.

B.5 Example of group of type curves

Fig. B.5 shows an example of the group of type curves.

B.6 Example of type curve matching method

Fig. B.6 shows an example of the type curve matching method.

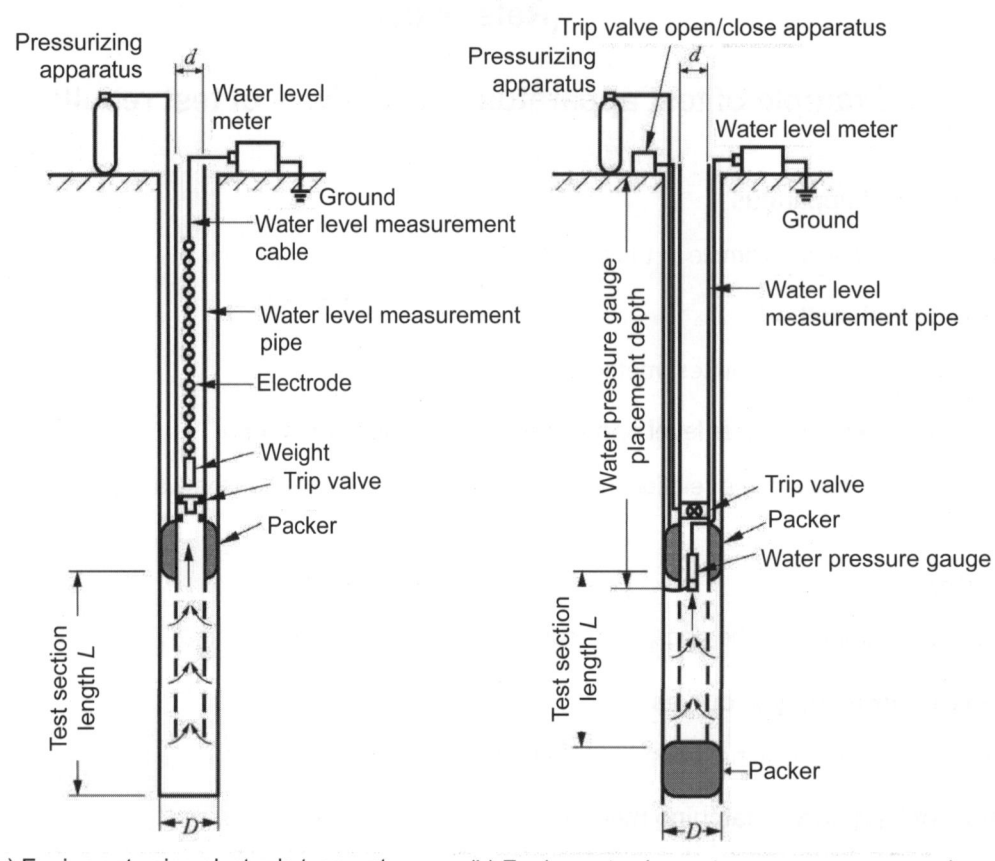

(a) Equipment using electrode type water level measurement cable (single packer)

(b) Equipment using water pressure gauge and open-close type trip valve (double packer)

Fig. B.1 Overview of test apparatus

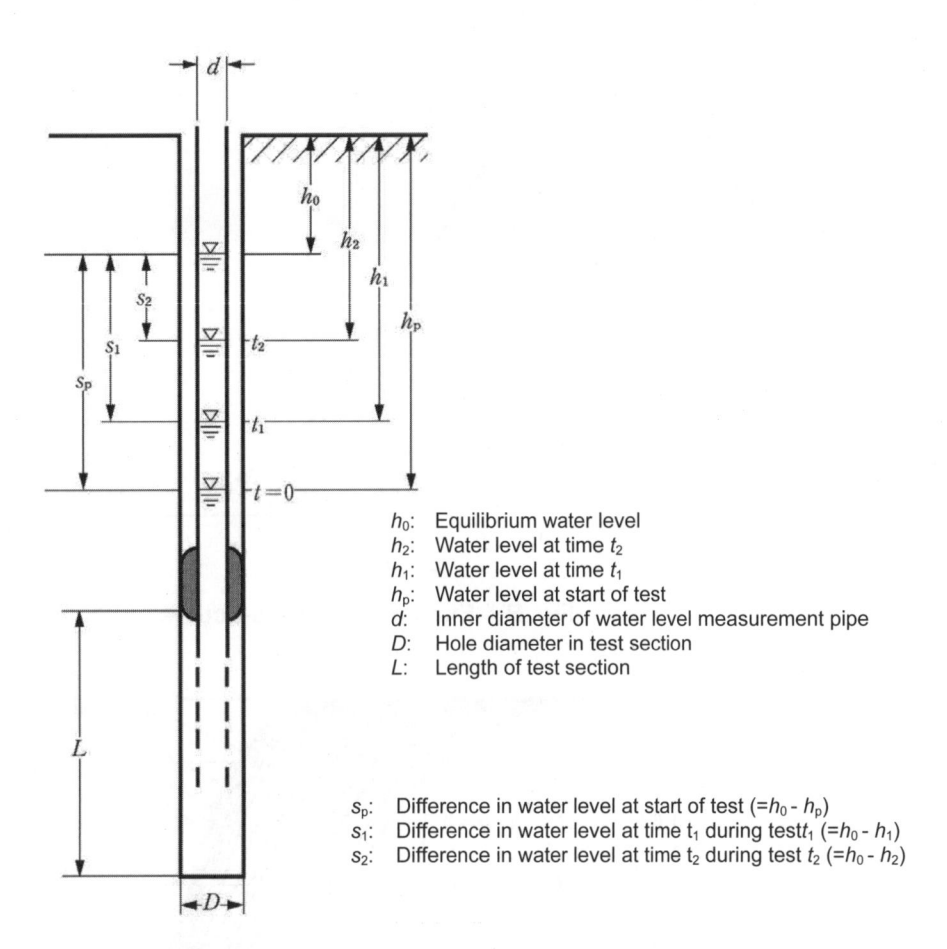

h_0: Equilibrium water level
h_2: Water level at time t_2
h_1: Water level at time t_1
h_p: Water level at start of test
d: Inner diameter of water level measurement pipe
D: Hole diameter in test section
L: Length of test section

s_p: Difference in water level at start of test (=$h_0 - h_p$)
s_1: Difference in water level at time t_1 during test t_1 (=$h_0 - h_1$)
s_2: Difference in water level at time t_2 during test t_2 (=$h_0 - h_2$)

Fig. B.2 Example of test method

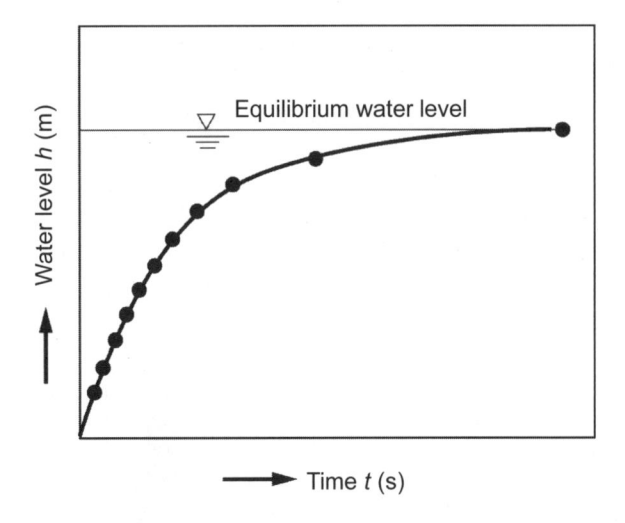

Fig. B.3 Example of relationship between water level and water level measurement time

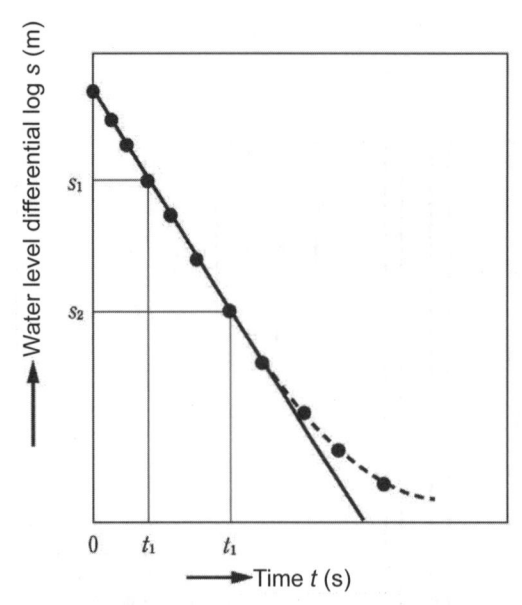

Fig. B.4 Example of $\log_{10}$ s-t curve

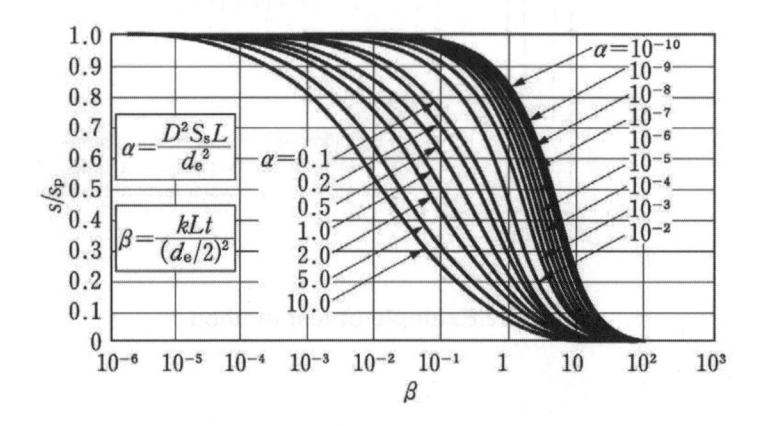

Fig. B.5 Example of group of type curves

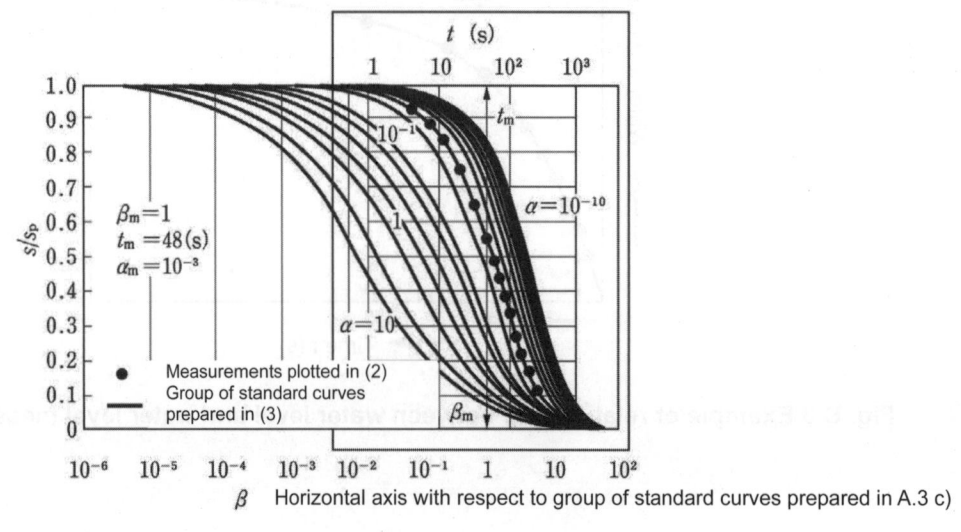

$$\alpha = \frac{D^2 S_s L}{d_e^2}$$

$$\beta = \frac{kLt}{(d_e/2)^2}$$

Horizontal axis with respect to measurements plotted in A.3 b)

$\beta_m = 1$
$t_m = 48 \text{ (s)}$
$\alpha_m = 10^{-3}$

- Measurements plotted in (2)
— Group of standard curves prepared in (3)

β Horizontal axis with respect to group of standard curves prepared in A.3 c)

Fig. B.6 Example of the type curve matching method

Japanese Geotechnical Society Standard (JGS 1322-2012)
Method for determination of hydraulic conductivity of rock mass using injection technique in single borehole

1 Scope

This standard specifies methods for determining the hydraulic conductivity of rock mass using a single borehole. This test applies to saturated rock mass beneath the level of groundwater.

Note 1: In this test, the hydraulic conductivity shall be determined by assuming that the rock mass is a homogenous porous medium.

2 Normative references

The following standard shall constitute a part of this standard by virtue of being referenced herein. The latest version of this standard shall apply (including supplements).

JIS B 7505-1   Aneroid Pressure Gauges - Part 1: Bourdon Tube Pressure Gauges

3 Terms and definitions

The main terms and definitions used in this standard are as follows:

3.1 Hydraulic conductivity test of rock mass using water injection

This is a steady-state hydraulic conductivity test in which a packer is used to demarcate an arbitrary test section in a borehole that has been drilled into a rock mass, after which water is injected into the test section while the effective water injection pressure is increased in stages, in order to determine the coefficient of hydraulic conductivity from the injection flow rate in the steady state. For this test of the hydraulic conductivity of rock mass, a low effective water injection pressure is used to ensure that no deformation or collapse of the rock mass occurs.

3.2 Effective water injection pressure

The effective water injection pressure is the water pressure during the injection minus the water pressure at equilibrium in the test section.

4 Equipment

Note: Fig. B.1 in Annex B shows an example of test equipment in which the water pressure gauge has been placed at ground level. Fig. B.2 in Annex B shows an example of test equipment in which the water pressure gauge has been installed in the test section.

4.1 Drilling equipment

The drilling equipment shall comprise a boring machine that can drill rock mass and clean the borehole.

4.2 Water injection pipe

The water injection pipe shall be a watertight pipe that can be used to measure the water level in the test section.

4.3 Packer

The packer shall adhere tightly to the borehole wall so that no water leakage is caused during the test.

Note 1: The types of packers comprise a single packer that is placed above the test section and a double packer that is placed both above and below the test section. The type shall be selected based on the condition of the borehole wall and the test procedure.

Note 2: The packer should be approximately 1 m or more in length.

4.4 Pressurizing apparatus

The pressurizing apparatus shall be able to inflate the packer using gas pressure or hydraulic pressure.

4.5 Water injection apparatus

The water injection apparatus shall be an airtight water tank or other equipment that can feed water by means of a pump or gas pressure.

Note: If a pump is used to inject water, an accumulator should be attached to reduce the pulsation of the water injection pressure.

4.6 Flowmeter

The flowmeter shall have a capacity that is appropriate for the water injection flow rate.

Note 1: The water injection flow rate shall depend on the hydraulic conductivity of the rock mass being tested. However, this is difficult to predict prior to the test, so a flowmeter capable of accommodating a wide range of water injection flow rates shall be prepared.

Note 2: If an airtight water tank is used as the water injection apparatus, the water level gauge in the water tank may serve as the flowmeter.

4.7 Water pressure gauge

The water pressure gauge shall be capable of measuring the water injection pressure in the test section.

Note 1: The water pressure gauge shall be placed at the borehole opening or in the test section. If the gauge is placed at the borehole opening, a water pressure gauge with a capacity approximately 1.5 times the maximum water injection pressure, and an accuracy within ±1.6% of the capacity, should be used. If the water pressure gauge is placed in the test section, a water pressure gauge with a capacity approximately 1.5 times the value of (unit weight of water x depth of test section) + maximum effective water injection pressure, and an accuracy within ±1.6% of the capacity, should be used. A water pressure gauge with an accuracy within ±1.6% of the capacity refers to a Bourdon tube pressure gauge with a 1.6 class accuracy classification as specified in JIS B 7505-1. If another water pressure gauge is used, the water pressure gauge should have an accuracy within ±1% of the capacity.

Note 2: The calibration coefficient and the indicator value of the water pressure gauge in no-load condition shall be checked in advance, and the water pressure gauge shall be calibrated prior to use.

4.8 Water level meter

The water level meter shall be capable of indicating the water level in the water injection pipe in 10 mm increments.

Note 1: If the water pressure gauge has been placed in the test section, the water pressure gauge may be used as an alternate for the water level meter.

5 Preparation of test hole

The test hole shall be prepared as follows.

a) Select a location for the test hole in an area that will not be affected by existing horizontal holes, boreholes and so on.

b) Use drilling equipment to drill the test hole.

Note 1: The test hole should be drilled using fresh-water drilling. However, if slurry is used, the borehole shall be flushed thoroughly

Note 2: In order to install the packer, water injection pipe and so on, and in order to maintain the shape of the test section, the finished test hole shall have stable borehole walls that will not collapse.

c) Flush the test hole until there is no slime or the like in the return water.

6 Test method

6.1 Test preparations

The following test preparations shall be made.

a) Determine the head loss h_3 (m) of the water injection pipe accompanying water injection by means of a head loss test [1] conducted on the ground.

Note [1]: A head loss test is a test conducted to determine the relationship between the water injection flow rate and the head loss of the water injection pipe.

Note 1: The head loss may be determined by calculation or through the use of a conversion table.

Note 2: The head loss test need not be performed if the water pressure gauge has been placed in the test section.

b) Determine the depth of the test section in accordance with the geological conditions of the rock mass being tested.

Note 3: The L/D ratio of the length L of the test section to the diameter D of the borehole shall be 4 or greater. If $L/D \geq 4$ cannot be secured, the test results shall be processed using a different equation from the one shown in A.2 in Annex A.

c) Measure the water level inside the test hole.

d) Attach the packer to the end of the water injection pipe.

e) While connecting the water level measurement pipe to which the packer has been attached, install the pipe to the depth determined in b).

Note 4: If the water pressure gauge is placed in the test section, record the depth at which the water pressure gauge has been placed.

f) Send fresh water through the water injection pipe to remove air bubbles from the interior of the pipe.

g) Using the pressurizing apparatus, inflate the packer so it adheres to the borehole wall.

Note 5: If the water shielding effectiveness of the packer is expected to be inadequate, protect the borehole walls with cement or the like in advance.

Note 6: If gas pressure is used to inflate the packer, apply sufficiently great gas pressure to prevent water leakage from occurring during the test, taking into consideration the fact that hydraulic pressure that is equal to the maximum effective water injection pressure will be applied to the test section.

Note 7: An effective way of confirming the water shielding effectiveness of the packer is to measure the water level inside the borehole above the packer (the water level in the gap between the water injection pipe and the borehole wall) before and after the test. If a significant increase in water level is observed after the test, this is thought to indicate that the water shielding effectiveness of the packer is inadequate. Change the position of the packer and conduct the test again.

Note 8: Measure the pressure of the gas or water used to pressurize the packer before and after the test. If there is a significant change in pressure after the test, there may be leakage from the packer or the piping system. In such cases, recover the packer and conduct inspection or replacement.

h) Using the water level meter, measure the water level in the water injection pipe at regular time intervals. Continue measuring until the change in water level becomes negligible. The value when the change in water level has become negligible is the equilibrium water level.

Note 9: If the water pressure gauge has been placed in the test section, this gauge may be used to measure the equilibrium water level.

Note 10: If the change in water level does not become negligible even after the equilibrium water level measurement has been conducted for a long period of time, estimate the equilibrium water level from the change trend. In such cases, the method used to estimate the equilibrium water level shall be reported as well.

6.2 Test

The test method shall be as follows.

a) Obtain the difference h_1 (m) between the height of the water pressure gauge and the height in the center of the test section, and the difference h_2 (m) between the equilibrium water level and the height in the center the test section.

b) Inject fresh water from the water injection apparatus to the test section at a constant water injection pressure p_1 (Pa).

c) Using the flow meter, measure the water injection flow rate once per minute to obtain readings for the change in water injection flow rate over time. The value when the water injection flow rate has become generally constant shall be the measured water injection flow rate Q_1 (m³/s) at water injection pressure p_1 (Pa).

d) Increase the water injection pressure and inject fresh water from the water injection apparatus to the test section at a constant water injection pressure p_i (Pa). Using the flow meter, measure the water injection flow rate once per minute to obtain readings for the change in water injection flow rate over time. The value when the water injection flow rate has become generally constant shall be the measured water injection flow rate Q_i (m³/s) at water injection pressure p_i (Pa). Here p_i (Pa) is the water injection pressure at stage i, and Q_i (m³/s) is the measured water injection flow rate at stage i.

Note 1: Instead of injecting water into the test section at a constant water injection pressure p_i (Pa), water may be injected into the test section at a constant water injection flow rate Q_i (m³/s). In such cases, the water injection pressure shall be measured once per minute using the water pressure gauge to obtain readings for the change in water injection pressure over time. The value when the water injection pressure has become generally constant shall be the measured water injection pressure p_i (Pa) at water injection flow rate Q_i (m³/s).

Note 2: If it is difficult to change the water injection pressure in stages and maintain a constant pressure, such as if the strength of the target rock mass is low or the hydraulic conductivity of the target section is high, the procedures in d) and e) may be omitted and the hydraulic conductivity may be determined from the water injection pressure p_i (Pa) and the measured water injection flow rate Q_i (m³/s) at a single stage. However, this method shall not be used if the equilibrium water level has not been measured or has not been suitably estimated.

e) Repeat d) for several water injection pressure stages. When the maximum water injection pressure stage has been completed, reduce the water injection pressure in stages and repeat the same measurements as in d).

Note 3: For the water injection pressure increase process, there shall be approximately five pressure stages, including the maximum water injection pressure. For the decrease process, there shall be approximately four pressure stages. Measurements shall be conducted during the water injection pressure increase process in order to calculate the hydraulic conductivity. Measurements shall be conducted during the pressure decrease process in order to judge whether or not clogging has occurred, whether or not a rock mass collapse has occurred and so on.

Note 4: At each pressure stage, the interval of time up until the pressure has been changed at the next pressure stage shall be made as constant as possible.

Note 5: The maximum water injection pressure shall be kept low so as to avoid crack deformation and rock mass collapse.

7 Analysis of test results

A method for analyzing the results is shown in Annex A.

8 Reporting

The following items shall be reported.

a) Number and location of test borehole, and ground elevation

Note 1: The elevation should be used for the ground level.

b) Depth, borehole diameter and length of test section

c) Geological conditions of test section

d) Test date and time, weather conditions and water level in borehole

e) Configuration of test apparatus

Note 2: The location of the water pressure gauge (depth or height from ground surface), the capacity and accuracy of the water pressure gauge, the length of the packer and the packer pressure shall be reported.

f) Flow rate measurement method

g) Equilibrium water level in test section

Note 3: If the change in water level did not become sufficiently small even after the equilibrium water level measurement was conducted for a long period of time, the method that was used to estimate the equilibrium water level shall also be reported together with the estimated equilibrium water level.

h) Time-based measurement records of water injection pressure and water injection flow rate

i) Relationship between effective water injection pressure head and measured water injection flow rate during the processes of water injection pressure increase and decrease

j) Maximum water injection pressure

k) Hydraulic conductivity

l) If the method used deviates in any way from this standard, give details of the method used.

m) Other reportable matters

Annex A
(Regulation)

Methods for analyzing results

A.1 Analysis

The results shall be analyzed as follows.

a) The effective water injection pressure head s_i (m) (the effective water injection pressure at each pressure stage converted into a water head value) shall be determined using the following equation.

$$s_i = \frac{p_i}{\gamma_w} + h_1 - h_2 - h_3$$

where

p_i: Water injection pressure at each pressure stage (Pa = N/m²)
h_1: Difference in height between water pressure gauge and center of test section (m)
h_2: Difference in height between equilibrium water level and center of test section (m)
h_3: Head loss due to resistance inside water injection pipe (m)

If the water pressure gauge has been placed inside the test section, $h_3 = 0$.

γ_w: Unit weight of water

Note 1: Regardless of whether the water pressure gauge has been placed at the borehole opening side or in the test section, the value indicated on the water pressure gauge shall be used as the water injection pressure p_i (Pa) at each pressure stage.

b) The measurement at each pressure stage shall be plotted as shown in Fig. B.3 in Annex B, with the effective water injection pressure head s_i (m) as the vertical axis and the measured water injection flow rate Q_i (m³/s) as the horizontal axis.

Note 2: The water injection pressure decrease process shall be used to determine whether or not clogging has occurred, whether or not a rock mass collapse has occurred as a result of water injection and so on. If the trend in the pressure decrease process differs significantly from the trend in the pressure increase process as shown in Fig. B.3 in Annex B, this shall be reported as a reportable matter.

A.2 Calculation of hydraulic conductivity

The hydraulic conductivity shall be calculated as follows.

a) From the drawing, determine the inclination a (s/m²) of the straight line section of the water injection pressure increase process. The inclination a (s/m²) shall be expressed using the following equation.

$$a = \frac{\Delta s}{\Delta Q}$$

Note 1: When determining the hydraulic conductivity from the water injection pressure p_1 (Pa) and measured water injection flow rate Q_1 (m³/s) at one stage, determine the inclination a (s/m²) using the following equation.

$$a = \frac{S_1}{Q_1}$$

b) Calculate the hydraulic conductivity k (m/s) of the rock mass using the following equation.

$$k = \frac{1}{2\pi L a} \ln\left(\frac{2L}{D}\right)$$

for

$$\frac{L}{D} \geq 4$$

where

    D: Hole diameter in test section (m)
    L: Length of test section (m)

Note 2: Use the same equation when determining the hydraulic conductivity from the water injection pressure p_1 (Pa) and measured water injection flow rate Q_1 (m³/s) at one stage.

Annex B
(Reference)

Example of test apparatus and analysis of test results

B.1 Example of test apparatus with water pressure gauge placed at ground level

Fig. B.1 shows an example of a test apparatus with the water pressure gauge placed at ground level.

B.2 Example of test apparatus with water pressure gauge placed in test section

Fig. B.2 shows an example of a test apparatus with the water pressure gauge placed in the test section.

B.3 Example of diagram showing relationship between effective water injection pressure head and measured water injection flow rate

Fig. B.3 shows an example of a diagram showing the relationship between the effective water injection pressure head and the measured water injection flow rate.

Note: The hydraulic conductivity shall be calculated from the pressure increase process. The pressure decrease process shall be used to determine whether or not clogging has occurred, whether or not a rock mass collapse has occurred as a result of water injection and so on. If the trend in the pressure decrease process differs significantly from the trend in the pressure increase process, this shall be reported as a reportable matter.

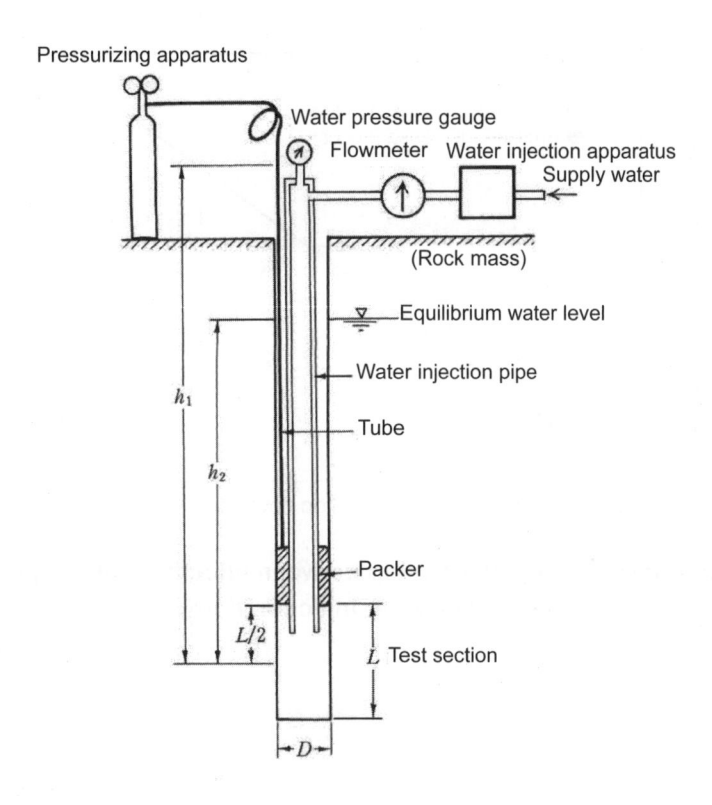

Fig. B.1 Example of test apparatus with water pressure gauge placed at ground level

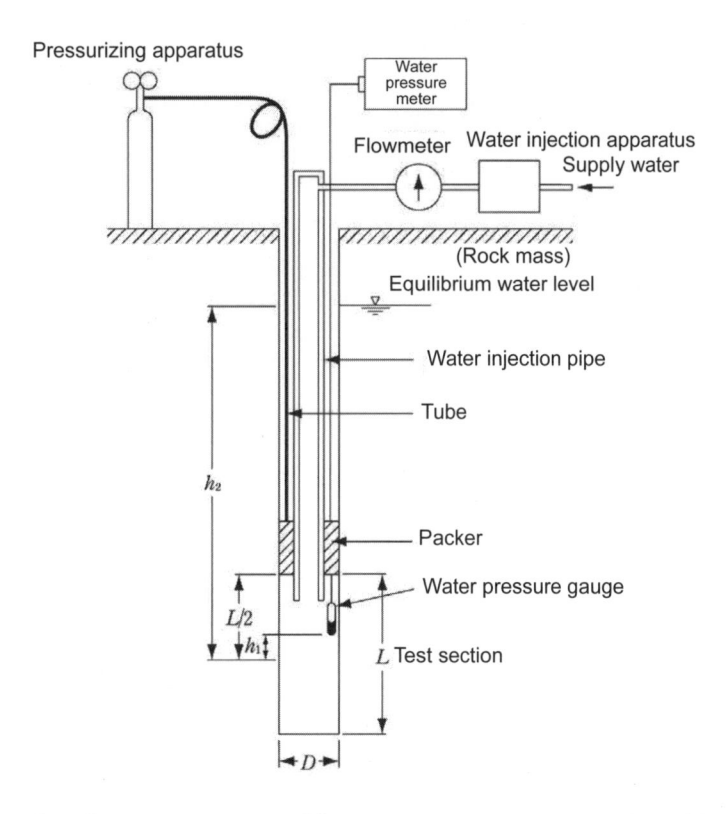

Fig. B.2 Example of test apparatus with water pressure gauge placed in test section

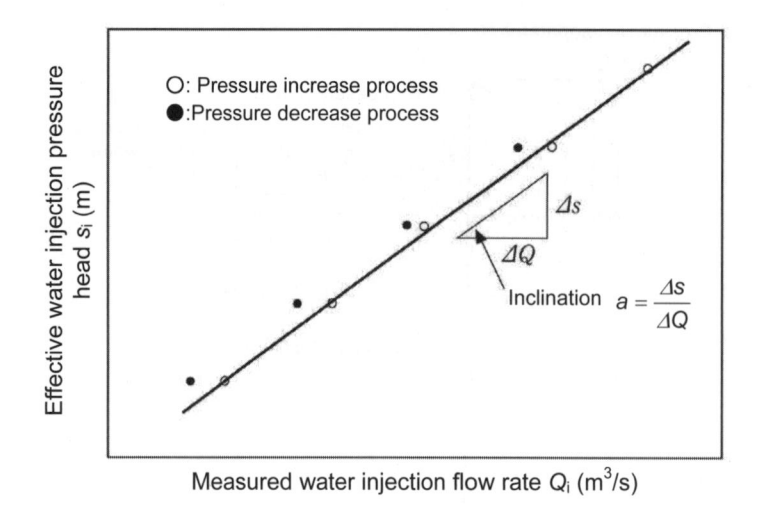

Fig. B.3 Example of diagram showing relationship between effective water injection pressure head and measured water injection flow rate

Japanese Geotechnical Society Standard (JGS 1323-2012)
Method for Lugeon Test

1 Scope

This standard specifies methods for determining the Lugeon value, an indicator of the hydraulic conductivity of rock mass. This test applies primarily to the rock mass of dam foundations.

Note: This test may also be applied to rock masses other than dam foundations.

2 Normative references

The following standard shall constitute a part of this standard by virtue of being referenced herein. The latest version of this standard shall apply (including supplements).

JIS B 7505-1 Aneroid Pressure Gauges - Part 1: Bourdon Tube Pressure Gauges

3 Terms and definitions

The main terms and definitions used in this standard are as follows:

3.1 Lugeon test

The Lugeon test is used to determine the Lugeon value by injecting water at a constant pressure into a test section inside a borehole that has been demarcated by a packer and then determining the relationship between the pressure and the water injection flow rate.

3.2 Lugeon value

The Lugeon value is the water injection quantity (ℓ) per minute per 1 m in a test section when water has been injected into the test section at an effective water injection pressure of 0.98 MPa.

3.3 Effective water injection pressure

The effective water injection pressure is pressure differential determined as the water pressure when water is injected into the test interval, minus the water pressure at equilibrium and a pressure equivalent to the head loss resulting from resistance inside the water injection pipe.

4 Equipment

Note: Fig. B.1 in Annex B shows an example of a Lugeon test apparatus.

4.1 Drilling equipment

The drilling equipment shall comprise a boring machine for rock mass that can drill and clean the interior of the borehole.

4.2 Water injection pipe

The water injection pipe shall be a pipe for injecting water into the test section, one that is watertight at water pressures of 2 MPa or greater. The connecting sections of the water injection pipe shall be provided with sealing tape or other means to prevent leakage.

Note 1: The inner diameter should be uniform.

4.3 Packer

The packer shall adhere tightly to the borehole wall and shall not produce water leakage during the test.

Note 1: The packer that is used should be an air packer that has excellent adhesion with the borehole wall.

Note 2: The types of packers comprise a single packer that is placed above the test section and a double packer that is placed both above and below the test section. The type shall be selected based on the condition of the rock mass and the test procedure.

Note 3: The packer should be approximately 1 m or more in length.

4.4 Pressurizing apparatus

The pressurizing apparatus shall be able to inflate the packer using gas pressure or hydraulic pressure.

4.5 Pump

The pump shall have an appropriate discharge pressure and discharge capacity and shall have low water injection pressure pulsation.

Note: The discharge capacity and maximum discharge pressure should be at least 150 ℓ/min (9 m^3/h) and 1.5 MPa, respectively, and the water injection pressure pulsation should be within ±10%.

4.6 Flow meter

The flow meter shall be able to measure the water injection flow rate.

Note 1: The flow meter should have a measurement accuracy within ±1.0% of the capacity.

Note 2: The flow meter should be able to conduct autographic recording. If the flow meter is not capable of autographic recording, as in the case of an integrating flow meter, accurate records shall be kept of the flow meter readings for each minute.

4.7 Pressure gauge

The pressure gauge shall be able to measure the water injection pressure in the test section.

Note 1: The pressure gauge shall be placed at the hole opening or in the test section and should be capable of autographic recording. If the gauge is placed at the hole opening, a pressure gauge with a capacity approximately 1.5 times the maximum water injection pressure, and an accuracy within ±1.6% of the capacity, should be used.

If the pressure gauge is placed in the test section, a pressure gauge with a capacity approximately 1.5 times the value of (unit weight of water x depth of test section) + maximum effective water injection pressure, and an accuracy within ±1.6% of the capacity, should be used. A water pressure gauge with an accuracy within ±1.6% of the capacity refers to a Bourdon tube pressure gauge with a 1.6 class accuracy classification as specified in JIS B 7505-1. If a different water pressure gauge is used, the water pressure gauge should have an accuracy within ±1% of the capacity.

Note 2: It is desirable to use a pressure gauge with an accuracy within 0.01 MPa if it is necessary to set the water injection pressure low and adjust the pressure step of the test in small increments.

4.8 Water level meter

The water level meter shall be capable of indicating the water level in the water injection pipe in 10 mm increments.

Note 1: If the water pressure gauge has been placed in the test section, the water pressure gauge may be used as the water level meter.

5 Preparation of test hole

The test hole shall be prepared as follows.

a) Select a location for the test hole in an area that will not be affected by existing horizontal shafts, boreholes and so on.

b) Use drilling equipment to drill the test hole and drill using fresh-water drilling. Normally the diameter of the test hole shall be 66 mm. However, if core sampling is conducted with a different hole diameter, that diameter may be used.

c) Flush the test hole until there is no slime or the like mixed in with the return water.

Note: In the case of soft rock, very weathered rock mass, significantly altered rock mass, a fragile fault fracture zone, an unconsolidated sediment layer or the like, flushing the interior of the borehole for long periods of time may itself disturb the borehole wall, so care shall be exercised.

6 Test method

6.1 Test preparations

The following test preparations shall be made.

a) Determine the head loss h_3 (m) of the water injection pipe accompanying water injection by means of a head loss test [1] conducted at ground level.

Note [1]: A head loss test is a test conducted to determine the relationship between the water injection flow rate and the head loss of the water injection pipe.

Note 1: The head loss may be determined by calculation or through the use of a conversion table.

Note 2: The head loss test need not be performed if the water pressure gauge has been placed in the test section.

b) Determine the depth of the test section in accordance with the geological conditions of the rock mass being tested. Normally the length of the test section shall be 5 m. However, the test section length may be reduced if necessary depending on the geological conditions and so on of the target section. In such cases, this fact shall be noted in the test results.

c) Measure the water level inside the test hole.

d) Attach the packer to the end of the water injection pipe.

e) While connecting the water injection pipe to which the packer has been attached, install the pipe down to the depth determined in b).

Note 3: If the pressure gauge is placed in the test section, record the depth at which the pressure gauge has been placed.

f) Send fresh water through the water injection pipe to remove air bubbles from the interior of the pipe.

g) Using the pressurizing apparatus, inflate the packer so it adheres to the borehole wall.

Note 4: If the water shielding effectiveness of the packer is expected to be inadequate, protect the borehole walls with cement or the like in advance.

Note 5: If gas pressure is used to inflate the packer, apply sufficiently great gas pressure to prevent water leakage from occurring during the test, taking into consideration the fact that hydraulic pressure that is equal to the maximum effective water injection pressure will be applied to the test section.

Note 6: An effective way of confirming the water shielding effectiveness of the packer is to measure the water level inside the borehole above the packer before and after water injection.

If a significant increase in water level is observed before or after water injection, this is thought to indicate that the water shielding effectiveness of the packer is inadequate, so change the position of the packer or take other steps.

Note 7: Measure the pressure of the gas or water used to pressurize the packer before and after the test. If there is a significant change in pressure before or after the test, there may be leakage from the packer or the piping system. In such cases, raise the packer and conduct inspection or replacement.

h) Using the water level meter, measure the water level in the water injection pipe at regular time intervals. Continue measuring until the change in water level becomes negligible. The value when the change in water level has become negligible is the equilibrium water level.

Note 8: If the water pressure gauge has been placed in the test section, this gauge may be used to determine the equilibrium water level.

Note 9: If the change in water level does not become negligible even after the equilibrium water level measurement has been conducted for a long period of time, estimate the equilibrium water level from the change trend. In such cases, the method used to estimate the equilibrium water level shall be reported as well.

6.2 Test

The test method shall be as follows.

a) Determine the difference h_1 (m) between the height of the water pressure gauge and the height in the center of the test section, and the difference h_2 (m) between the equilibrium water level and the height in the center the test section.

b) Operate the pump to inject fresh water into the test section at a constant water injection pressure p_0 (MPa).

c) Using the flow meter, measure the water injection flow rate once per minute to obtain readings for the change in water injection flow rate over time. When the range of fluctuations in the water injection flow rate has become less than 10% of the average water injection flow rate over a period of five minutes, this average water injection flow rate shall be used as the measured water injection flow rate Q_0 (ℓ/min) at the water injection pressure p_0 (MPa).

d) Increase the water injection pressure in stages from 0 and conduct the test. After the test at the maximum water injection pressure stage has been completed, reduce the pressure in stages to 0. During this process, repeat the measurements in b) and c). For the process of increasing the water injection pressure, there shall be five or more pressure stages, including the maximum water injection pressure. For the decrease process, there shall be four or more pressure stages. Normally the maximum effective water injection pressure shall be 0.98 MPa. However, for ground whose limit pressure is less than 0.98 MPa, short intervals shall be used for the pressure increase and decrease stages so the pressure increase is conducted in five or more stages.

Note: The same pressure values should be used for the pressure increase stages and pressure decrease stages. An example of the water injection pressure pattern is shown in Fig. B.2 in Annex B.

7 Analysis of test results

Plot a graph with the effective water injection pressure p at each pressure stage as the vertical axis and the water injection flow rate q per unit of length as the horizontal axis, and determine the Lugeon value from the relationship between p and q.

A method for analyzing the results is shown in Annex A.

8 Reporting

The following items shall be reported.

a) Number and location of test borehole and ground elevation, and drilling method

Note 1: The elevation should be determined as the ground height.

b) Depth, borehole diameter and length of test section

c) Geological conditions of test section

d) Test date and time and weather conditions

e) Configuration of test apparatus

Note 2: The location of the pressure gauge (depth or height from ground surface), the capacity and accuracy of the pressure gauge, the length of the packer and the packer pressure shall be reported.

f) Equilibrium water level in test section

Note 3: If the change in water level did not become sufficiently small even after the equilibrium water level measurement was conducted for a long period of time, the method that was used to estimate the equilibrium water level shall also be reported together with the estimated equilibrium water level.

g) Time-based measurement records of water injection pressure and water injection flow rate

h) Relationship between effective water injection pressure and water injection flow rate per unit of length during the processes of water injection pressure increase and decrease ($p-q$ line)

i) Maximum water injection pressure

j) Lugeon value (L_u) or converted Lugeon value (L_u') and limit pressure

k) If the method used deviates in any way from this standard, give details of the method used.

l) Other reportable matters

Annex A
(Regulation)

Methods for analyzing results

A.1 Analysis of test results

The results shall be analyzed as follows.

a) Determine the effective water injection pressure p (MPa) at each pressure stage, using the following equation.

$$p = p_0 + \gamma_w(h_1 - h_2 - h_3)$$

where

p_0: Water injection pressure at each pressure stage (MPa)
h_1: Difference in height between water pressure gauge and center of test section (m)
h_2: Difference in height between equilibrium water level and center of test section (m)
h_3: Head loss due to resistance inside water injection pipe (m)

If the water pressure gauge has been placed inside the test section, $h_3 = 0$.

γ_w: Unit weight of water (MN/m^3)

Note: Regardless of whether the water pressure gauge has been placed at the borehole opening side or in the test section, the value indicated on the water pressure gauge shall be used as the water injection pressure p_0 at each pressure stage.

b) Calculate the water injection flow rate q ($\ell/(\min \cdot m)$) for each unit of length, using the following equation.

$$q = \frac{Q_0}{L}$$

where

Q_0: Measured water injection flow rate ($\ell/\min$)
L: Length of test section (m)

c) Plot a graph with the effective water injection pressure p (MPa) as the vertical axis and the water injection flow rate q ($\ell/(\min \cdot m)$) per unit of length as the horizontal axis, and plot the measurements in order to draw a $p-q$ line as shown in Fig. B.3 in Annex B.

A.2 Method for determining Lugeon value

The method for determining the Lugeon value shall be as follows.

a) From the linear relationship of the $p-q$ line, determine the water injection flow rate q ($\ell/(\min \cdot m)$) for each unit of length that corresponds to an effective water injection pressure of 0.98 MPa, and use this value as the Lugeon value L_u.

b) If any sudden bends such as that shown as in Fig. B.4 in Annex B are produced on the $p-q$ line, use the effective water injection pressure at the intersection between the two straight lines as the limit pressure p_{cr} (MPa).

c) If the limit pressure or the maximum water injection pressure is less than 0.98 MPa, extend the initial straight line section during the pressure increase process in Fig. B.4 in Annex B and determine the water injection flow rate $q\ (\ell/(\min\cdot m))$ for each unit of length that corresponds to an effective water injection pressure of 0.98 MPa, and use this value as the converted Lugeon value L_u'.

Annex B
(Reference)

Example of test apparatus, water injection pressure pattern, and analysis of test results

B.1 Example of Lugeon test apparatus

Fig. B.1 shows an example of the Lugeon test apparatus.

B.2 Example of water injection pressure pattern

Fig. B.2 shows an example of the water injection pressure pattern.

B.3 Example of method for determining Lugeon value

Fig. B.3 show an example of the method used to determine the Lugeon value.

B.4 Example of method for determining converted Lugeon value

Fig. B.4 shows an example of the method used to determine the converted Lugeon value.

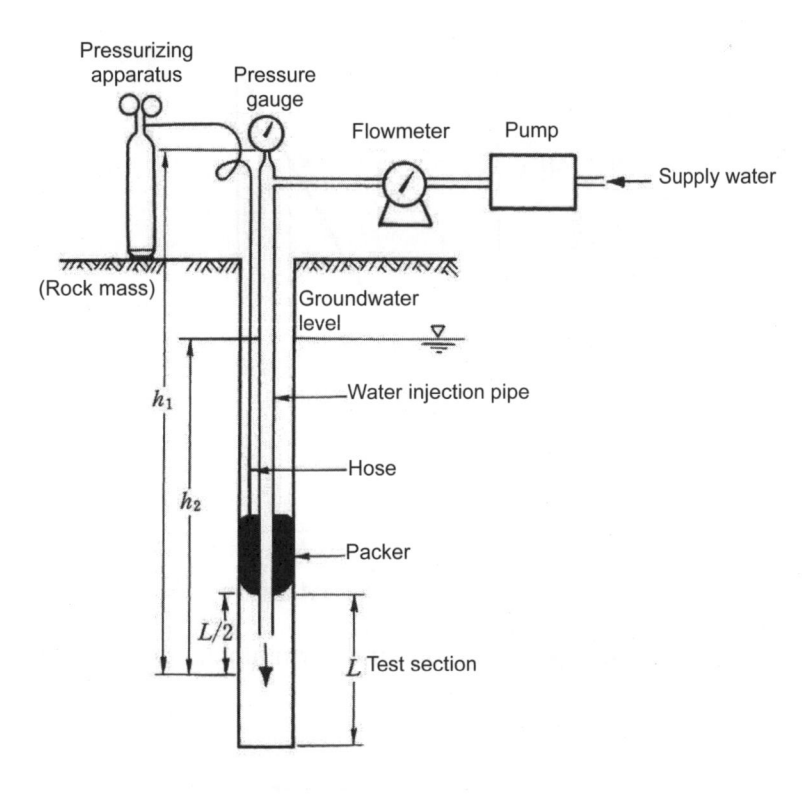

Fig. B.1 Example of Lugeon test apparatus

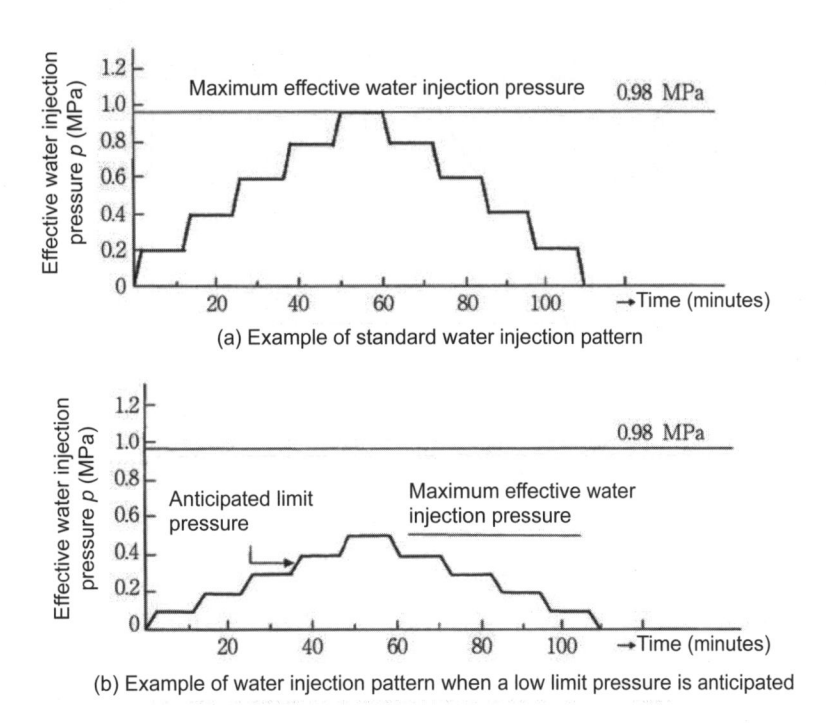

(a) Example of standard water injection pattern

(b) Example of water injection pattern when a low limit pressure is anticipated

Fig. B.2 Example of water injection pressure pattern

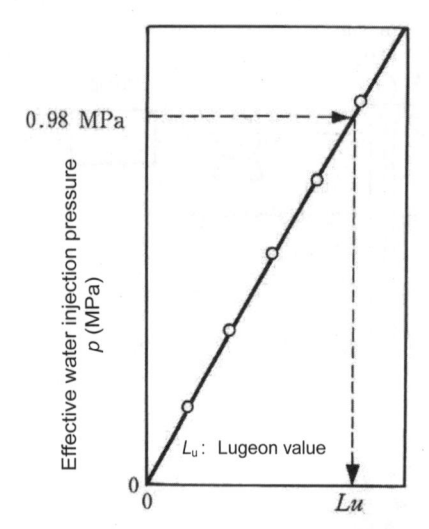

Fig. B.3 Example of method for determining Lugeon value

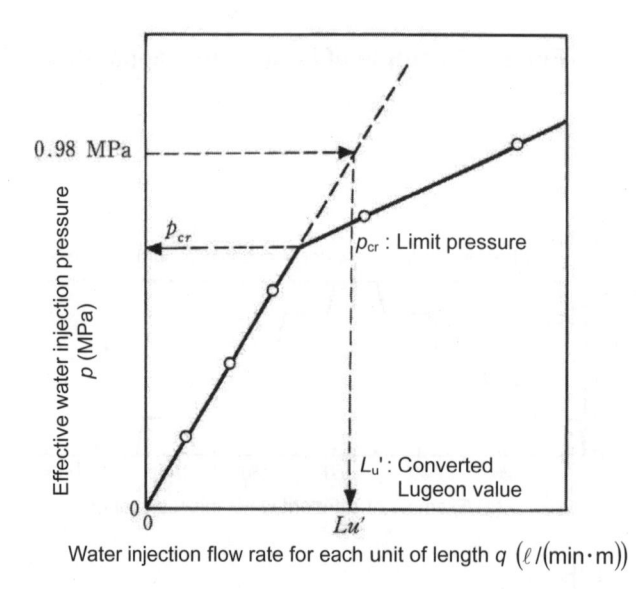

Fig. B.4 Example of method for determining converted Lugeon value

Japanese Geotechnical Society Standard (JGS 1437-2014) Method for dynamic cone penetration test

1 Scope

This document covers the determination of the resistance of soils and soft rocks in situ to the dynamic penetration of a cone.

A hammer of a given mass and given height of fall is used to drive the cone. The penetration resistance is defined as the number of blows, N_d, required to drive the cone over a defined distance. A continuous record is provided with respect to depth but no samples are recovered. N_d can be used to estimate rigidity and degree of compaction of soils.

This document is applied to medium and heavy types of dynamic cone penetration test (hereafter, DCPT) having different values of specific work per blow as follows:

   Medium DCPT: DCPT with specific work per blow equal to 98 kJ/m^2.

   Heavy DCPT: DCPT with specific work per blow equal to 196 kJ/m^2.

Note: Medium DCPT is equivalent to DPM (medium) and heavy DCPT is equivalent to DPSH (super heavy-A) specified in ISO 22476-2 Geotechnical investigation and testing -Field testing- Part 2: Dynamic probing, 2005.

The test results of this document are suited for the determination of the depth of a bearing stratum. They are also used for detecting very loose soils, cavities, fills and refilled soils. They may also be used for the determination of the soil properties together with direct investigations (e.g. soil sampling).

2 Normative references

None

3 Terms and definitions

For the purpose of this document, the following terms and definitions apply.

3.1 Dynamic cone penetration test

Test to drive a cone fitted to the end of rods into the ground with a specific work per blow

3.2 Drive-weight assembly

Device consisting of a hammer, a hammer fall guide, an anvil, and an automatic drop system

3.3 Hammer

Portion of the drive-weight assembly that is successively lifted and dropped to provide the energy that accomplishes the penetration of the cone

3.4 Hammer fall guide

Portion of the drive-weight assembly that guides the free fall of the hammer to the anvil

3.5 Anvil

Portion of the drive-weight assembly that the hammer strikes and through which the hammer energy passes into the drive rods

3.6 Cushion; Damper

Material placed upon the anvil to minimize noise and damage to the equipment during testing

3.7 Automatic drive-weight assembly

Device with the function of automatically guiding the free fall of the hammer.

3.8 Height of fall

Height of free fall of the hammer after being released.

3.9 Drive rods

Rods that connect the anvil to the cone.

3.10 Cone

Pointed conical probe and its cylindrical extension (mantle) (see Figure 1).

3.11 N_d-value

Number of blows required to drive the cone the defined distance of 200 mm.

Note 1: As the specific work per blow is different between the medium and heavy DCPTs, the resulting N_d-values are different.

Note 2: The N_d-value specified in JGS 1433 (Method for Portable Dynamic Cone Penetration Test) is number of blows required to drive the cone the defined distance of 100 mm. As specific work per blow of the portable dynamic cone penetration test is different from that of the medium DCPT or the heavy DCPT, the N_d-value obtained in the portable dynamic cone penetration test is different from those obtained in the medium DCPT or the heavy DCPT.

3.12 Specific work per blow E_n

Value calculated by

$$E_n = m \times g_n \times h / A$$

where

- m: the mass of the hammer;
- g_n: the standard acceleration due to gravity;
- h: the height of fall of the hammer;
- A: the nominal base area (calculated using the base diameter D)

3.13 Torque M_v

The maximum torque measured when the rods are rotated, for measuring the skin friction acting on the rods in the ground.

3.14 Automatic measuring and recording device

Device consisting of sensors for measuring number of blows of the hammer, the penetration length, the torque, etc., and recording system.

3.15 Penetration length

Total length of the cone and the rods that have penetrated into the ground.

Note: The distance between the cone base to the cone tip is not included in the penetration length. The penetration length does not necessarily coincide to the depth of cone base when verticality of the rods is not maintained.

4 Equipment

An example of dynamic cone penetration test equipment is shown in Annex A.

4.1 Driving device

Dimensions and masses of the components of the driving device are given in Table 1. The following requirements shall be fulfilled.

a) The hammer shall be guided by the hammer guide to ensure minimal resistance during the fall.

b) The automatic release mechanism shall ensure that the hammer is virtually static at the time of being released, and the prescribed height of fall can be provided without causing oscillations, etc., in the rods.

c) An anvil that is rigidly connected to the top of the drive rods or an anvil that is non-rigidly connected to the top of the drive rods shall be used.

4.2 Anvil

The anvil shall be made of high strength steel. A damper or cushion may be fitted between the hammer and anvil, to reduce impact noise of the hammer or to reduce damage to the equipment during testing.

4.3 Cone

The steel cone shall have the dimensions shown in Figure 1 and Table 1. The cone shall have an apex angle of 90 degrees and an upper cylindrical extension mantle. The cone may be either a recovery or disposable type, but these types shall have the same dimensions.

Note: The recovery type cone can be recovered when the rods are extracted after the test. It is recommended that a mechanism be provided so that when the rods are rotated during torque measurement, the rods only rotate, and the cone does not rotate. The disposal type cone separates from the rods when the rods are withdrawn after the test, and remains in the ground. When using a disposable cone the end of the drive rod shall fit tightly into the cone. Alternative specifications for the cones are given in Figure 1.

4.4 Drive rods

The rod shall have dimensions and masses given in Table 1, and shall be made from a high-strength steel with the appropriate characteristics for the work to be performed without excessive deformations and wear.

The rods shall be flush jointed, shall be straight and may have spanner flats. The deflection at the mid point of an extension rod measured from a straight line through the ends shall not exceed 1 in 1000 (0.1%), i.e. 1 mm in 1 m.

4.5 Torque measuring device

The torque necessary to turn the driving rods for measuring the skin friction between the rods and ground is measured by means of a torque wrench or similar measuring device. Spanner flats may be used on the rods when using a torque wrench. The apparatus shall be capable of measuring a torque of at least 100 Nm or 200 Nm and shall be graduated to read in increments of 5 Nm or 10 Nm or less, for the medium and the heavy types of DCPT, respectively.

A sensor for automatically recording the torque may be used.

4.6 Optional equipment

4.6.1 Automatic measuring device

A device to count the number of blows of the hammer by measuring mechanical or electric impulses can be placed on the system. The penetration length may be measured using a sensor that shall have a resolution of 2 mm or better.

4.6.2 Device to maintain verticality of the rod string

A system or guide for supporting the protruding part of the rods should be in place to ensure and check that the drive rods are maintained in vertical alignment.

4.6.3 Rod extraction device

A rod extraction device is used to extract the rods after the test, and shall have sufficient capacity for extraction.

5 Test procedure

5.1 Selection of DCPT device

The test shall be carried out with the appropriate selection of either of the medium or the heavy DCPT apparatus, according to soil conditions, site conditions, and purpose of the test.

5.2 Equipment checks and calibrations

Prior to each test, a check of dimensions and masses shall be made to ensure that they are within the values given in Table 1. The straightness of the rods shall be visually checked once on each new site and at least once every 20 penetration tests at that site. At the test site, the number of blows per minute, the height of fall, the flatness of the bottom surface of the hammer and the anvil pressure surface, and the mechanical release device shall be checked, in order to ensure proper operation of the test equipment. In addition, the proper functioning of the recording device shall to be checked if automatic recording equipment is used.

The precision of the automatic measuring device -if applicable- shall be checked after any damage, overloading, or repair, and at least once every six months Faulty parts shall be replaced. Calibration records shall be kept together with the equipment.

5.3 Test preparation

In general, DCPT is performed from the ground surface.

DCPT equipment shall be set up with the penetrometer vertical, and in such a way that there will be no horizontal displacement during testing. The inclination of the driving mechanism and the driving rod projecting from the ground shall be not more than 2% from the vertical.

Trailer-mounted DCPT equipment shall be supported with outriggers in such a way that the suspensions of the support trailer cannot affect the test.

The equipment shall be set up with appropriate clearance from structures, piles, boreholes etc., in order to be certain that they will not affect the result of the DCPT.

When carrying out DCPT in situations where the rods are free to move laterally, for instance over water or in boreholes, the rods shall be restrained by low-friction supports in order to prevent bending during driving.

5.4 Test execution

The drive rods and the cone shall be driven vertically and without undue bending of the protruding part of the extension rods above the ground. When a rod is added, the verticality of the rod should be less than 2% from the vertical. If the verticality of the rods exceeds 5%, this situation shall be reported.

The penetrometer shall be continuously driven into the ground. The driving rate shall be kept between 15 and 30 blows per minute. All interruptions longer than 5 minutes shall be recorded.

The number of blows, N_d, shall be recorded every 200 mm penetration.

In cases of DCPT where the driving resistance is very low, such as in soft clays, the penetration distance per blow may be recorded.

The penetration distance of the cone due to the self-weight of the equipment shall be recorded, if it occurs.

Note: If the cone penetration due to the self-weight of the equipment occurs during impacts, the next blow shall be carried out after it has been confirmed that the penetration due to the self-weight has virtually ceased.

The rods shall be rotated 1.5 turns or more or until maximum torque is reached at least every 1.0 m penetration. The maximum torque required to turn the rods shall be measured using a torque measuring wrench or an equivalent device and shall be recorded. However after torque measurement during heavy driving, the rods shall be rotated 1.5 turns or more after every 50 blows to tighten the rod connections.

If the penetration distance does not reach 200 mm after 200 blows for the medium DCPT and after 100 blows for the heavy DCPT, the test may be terminated. The penetration distance at the termination shall be recorded. In cases where N_d greater than 100 for the medium DCPT and N_d greater than 50 for the heavy DCPT is successively recorded 5 times, the test may be terminated.

After the termination of the test, driving equipment shall be dismounted and the rods shall be extracted using rod extraction device. Visual inspection of the rods shall be made after the test.

5.5 Safety requirements

National safety regulations shall be followed.

6 Test results

The test results shall be reported as values of the number of blows N_d and the torque M_v with penetration length.

7 Reporting

7.1 Field report

7.1.1 General

At the project site, a field report shall be completed. This field report shall include records of the measured values and test results, if applicable.

All field reports shall be recorded so that third parties can check and understand the results.

7.1.2 Record of measured values and test results

At the project site, the following information shall be recorded for each test. Note that items with * symbol shall be included. The other items may be recorded, if required.

a) General information:

    1) Name of the client;

    2) Name of the contractor;

    3) Job or site number;

    4) Name and location of the site;

    5)* Name of the test equipment operator in charge;

b) Information on the test position:

    1)* Date and number of test;

    2) Field sketch (scale does not matter);

3)* Ground elevation referred to a reference point;

4) Location in plan of the test position;

5) Operation on land or water;

c) Information on the test equipment used:

1)* Type of dynamic cone penetration test (medium or heavy);

2) Manufacturer, model, and serial number of the test equipment;

3)* type of cone (disposable or fixed);

4) Type of anvil (fixed or loose connection to rod);

5) Use of dampers or cushions, and their material and dimensions;

d) Information on the test procedure:

1) Weather conditions;

2) Documentation of the equipment check and calibration conducted in accordance with 5.2;

3)* The following test results:

- N_d-value with penetration length;
- M_v-value with penetration length;
- Penetration distance per blow, if applicable;
- Penetration distance, if the test was terminated before the penetration distance reached 200 mm;
- Penetration distance due to the self-weight of the equipment, if it occurred;

4) Blow count per minute;

5) Groundwater level, artesian conditions, if measured;

6)* All unusual events or observations during the operation (e.g. malfunction of the equipment);

7) Observations on the recovered cone and/or rods;

8)* All interruptions during the work

9)* Reasons for early termination of the test;

10)* Observations when the verticality of the rods exceeded 5 % (e.g. penetration length, reasons).

7.2 Test report

For checking the quality of the data, the test report shall include the following in addition to the information given in 7.1:

a) Field report (print out);

b) Graphical representation with respect to depth of the following data:

- Graph of number of blows N_d on the horizontal axis and penetration length on the vertical axis;
- Graph of the maximum torque M_v on the horizontal axis and penetration length on the vertical axis;

c) If a method that partially differs from this standard was used, details of the points of difference ;

d) Name the field manager.

The test results shall be reported in such a manner that a third person is able to check and understand the results.

Table 1 Dimensions and masses for the medium and heavy types of dynamic cone penetration testing apparatus

Dynamic Cone Penetration Test Apparatus	Symbol	Unit	Medium	Heavy
Driving device				
Hammer mass	m	kg	30.0±0.3	63.5±0.5
Height of fall	h	mm	350±10	500±10
Maximum total mass [a]		kg	70	115
Anvil				
Diameter [b]	d	mm	$50 < d < D_h$	$50 < d < 0.5 D_h$
Mass (max.) [c]	m	kg	18	18
Cone				
Nominal base area	A	mm²	1 052	1 590
Base diameter	D	mm	36.6±2.0	45.0±2.0
Mantle length	L	mm	69.0±3.0	90.0±2.0
Length of cone tip	L_c	mm	18.3±2.0	22.5±2.0
Drive rods				
Mass (max.)	m	kg/m	5.0	6.7
Diameter OD	d_r	mm	28.0±0.4	32.0±0.4
Rod deviation		%	0.1	0.1

Note [a] The mass that acts on the rods after the blow.
Note [b] D_h is the outer diameter of the hammer.
Note [c] Guide rod included.

(a) recovery type

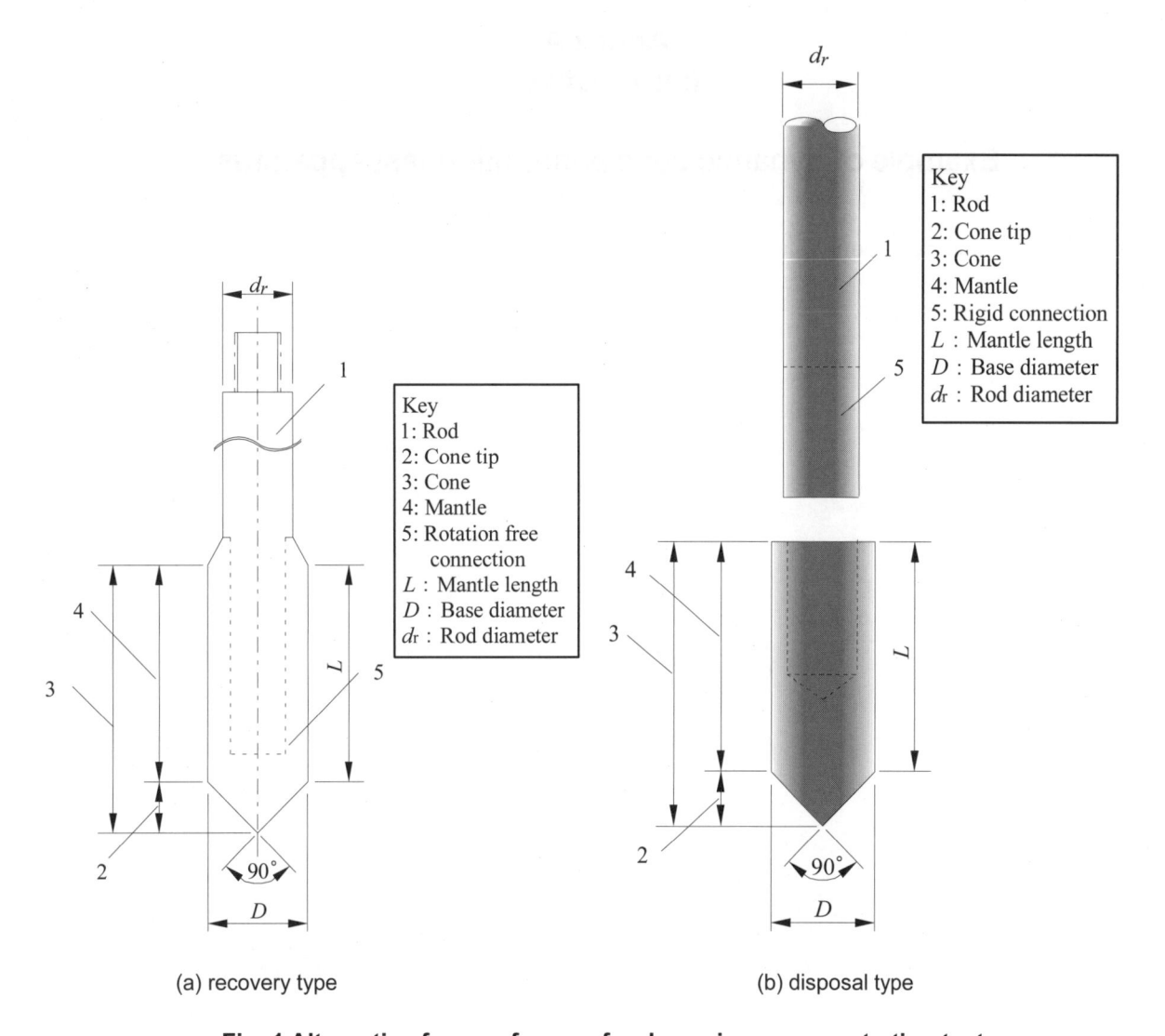

Key
1: Rod
2: Cone tip
3: Cone
4: Mantle
5: Rotation free connection
L : Mantle length
D : Base diameter
d_r : Rod diameter

(b) disposal type

Key
1: Rod
2: Cone tip
3: Cone
4: Mantle
5: Rigid connection
L : Mantle length
D : Base diameter
d_r : Rod diameter

Fig. 1 Alternative forms of cones for dynamic cone penetration test

Annex A
(Informative)

Example of dynamic cone penetration test apparatus.

An example of dynamic cone penetration test apparatus is shown in Figure A.1.

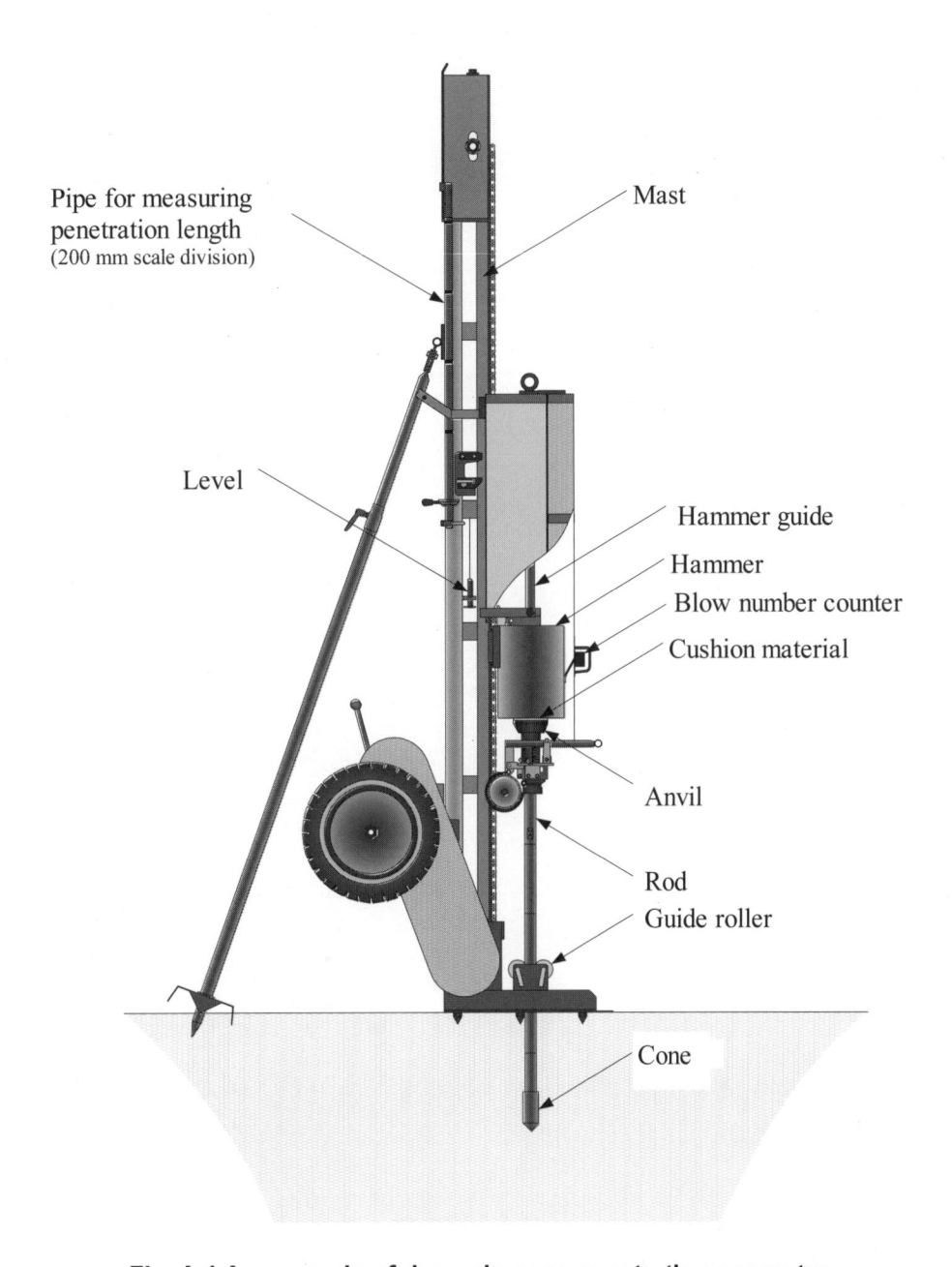

Fig. A.1 An example of dynamic cone penetration apparatus

Japanese Geotechnical Society Standard (JGS 1441-2012)
Method for soil hardness test

1 Scope

This standard specifies methods for measuring the soil hardness of the surface of natural slopes, cut earth slopes, tunnel faces and other exposed ground as well as penetrable solidification-treated soil, improved soil specimens, boring samples and so on, using a penetration type soil hardness meter.

2 Normative references

None

3 Terms and definitions

The main terms and definitions used in this standard are as follows:

3.1 Cone

The cone is a conical part mounted on the end of a penetration type soil hardness meter.

3.2 Penetration length

The penetration length is the length to which the cone penetrates the soil.

Note: The penetration length shall be expressed in mm.

3.3 Indicated hardness

The Indicated hardness is the hardness of soil expressed as the penetration length of a penetration type soil hardness meter. The penetration length is determined by reading the length to which the member that is rigidly connected to the cone is pressed into the cylinder that follows behind.

3.4 Indicated hardness scale

The indicated hardness scale is the scale on which the penetration length of the penetration type soil hardness meter is indicated.

3.5 Movement indicator

The movement indicator is the indicator that shows the penetration length of the penetration type soil hardness meter.

3.6 Measurement surface

The measurement surface is the shaped surface that is used to measure the soil hardness.

4 Equipment

The penetration type soil hardness meter that is used for the test shall fulfill the following requirements.

4.1 Shape

The soil hardness meter shall be one with a conical end as shown in Fig. 1, and the amount of penetration of the conical-shaped cone shall be indicated in mm units.

4.2 Cone

The cone is the section that penetrates the soil. It shall be conical in shape, with a length of $40^{0}_{-0.5}$ mm and a base diameter ø of $18^{0}_{-0.2}$ mm, and shall be made of hardened, wear-resistant and rust-resistant steel. The surface shall be smooth with no scratches or the like. The end shall be sharp with no rounding due to wear, and it shall not be broken or bent. The material shall be stainless steel (SUS 304, JIS G 4303).

4.3 Abutment flange

This is the section that contacts the measurement surface when the cone penetrates the soil. It is a circular ring with diameter of (38±0.5) mm, thickness (3±0.2) mm and hole diameter (18+0.2) mm.

4.4 Spring

The spring is a coil spring attached to the interior and connected to the cone. The spring contracts proportionally to the load as the cone penetrates the soil. The load shall be (78.4±2.0) N for a contraction quantity of 40 mm.

4.5 Movement indicator

The movement indicator moves along with the contraction of the spring due to the cone penetration load, and it stops at an arbitrary point to accurately indicate the measurement. When the cone penetrates the soil, the rigidly connected member is pressed into the cylindrical section. As this happens, the movement indicator moves along the indicator scale, and a mechanism causes it to stop at the position to which it moved in response to cone penetration, even after the cone has been removed from the soil.

4.6 Indicated hardness scale

The indicated hardness scale is used to indicate the contraction length of the spring that corresponds to the degree to which the section that follows is pressed inside the cylinder by the cone penetration load. Markings in mm increments 0 - 40 mm are displayed on the cylindrical section of the penetration type soil hardness meter, with minimum scale markings for each 1 mm, medium-sized scale markings for every 5 mm and large scale markings for every 10 mm. The numbers 0, 10, 20, 30 and 40 are stamped at the large scale markings.

5 Test method

The test method for in-situ measurement shall comprise, in order, the following steps: shaping of the measurement surface, cone penetration, cone removal and reading of the indicated hardness, and finalization of the indicated hardness.

5.1 Shaping of measurement surface

The measuring surface shall be dug approximately 10 mm into the slope to be measured, as shown in Fig. 2, to create a measurement surface by shaping a flat surface that measures approximately 300 mm square. When the measuring surface is moist with rain or the like, it shall be dug down to where it is judged to show the original moisture condition.

5.2 Cone penetration

Set the movement indicator to 0 on the indicated hardness scale. Place the cone end vertically against the measurement surface and push it gradually into the soil until the abutment flange contacts the measurement surface completely. During this process, the scale shall be facing to the side or downward to ensure that no soil gets into the slit of the movement indicator.

When penetration is stopped, there should be no gap between the abutment flange and the measurement surface.

Note 1: If the cone has clearly come into contact with gravel or the like during cone penetration, avoid that location and repeat the previous procedure.

Note 2: The penetration speed will affect the soil resistance. The period of time needed for penetration shall be approximately two seconds.

5.3 Cone removal and reading of indicated hardness

Remove the cone gently, making sure the movement indicator does not move, and record the reading indicated by the movement indicator on the indicator hardness scale. Exercise care during this process, as the movement indicator may be moved if the cone is removed too quickly.

5.4 Repeated measurement

At least five measurement points shall be set on the measurement surface as shown in Fig. 2 and repeated measurement is performed. During this process, carefully remove any soil sticking to the cone or soil that has gotten into the slit on the movement indicator section, and then return the movement indicator to 0 on the indicator hardness scale and repeat the procedure from 5.2.

Note: If the area of the measurement surface is small, as in the case of a boring sample or improved soil specimen, the number of measurement points may be reduced and a representative value may be used.

5.5 Determination of indicated hardness

Of the five or more indicated hardness values recorded for a single measurement surface, discard the highest and lowest values and determine the average of the remaining measurements. Use this value as the indicated hardness of that measurement surface.

6 Reporting

The following items shall be reported as the test results.

a) Purpose of test

b) Measurement number

c) Test position

d) Indicated hardness (individual values and average value, in mm)

e) Other reportable matters

Unit: mm

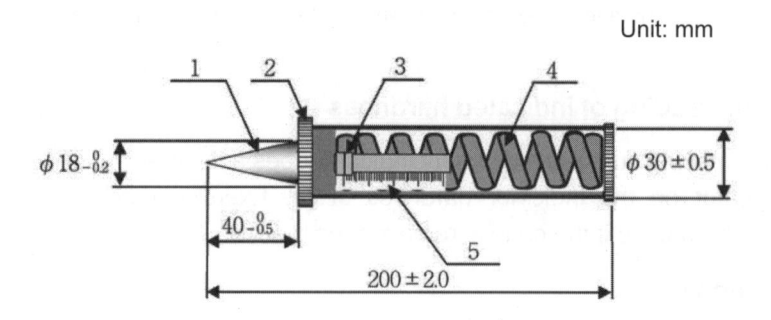

1 Cone
2 Abutment flange
3 Movement indicator
4 Spring
5 Indicated hardness scale

Fig. 1 Penetration type soil hardness meter

○ Measurement point

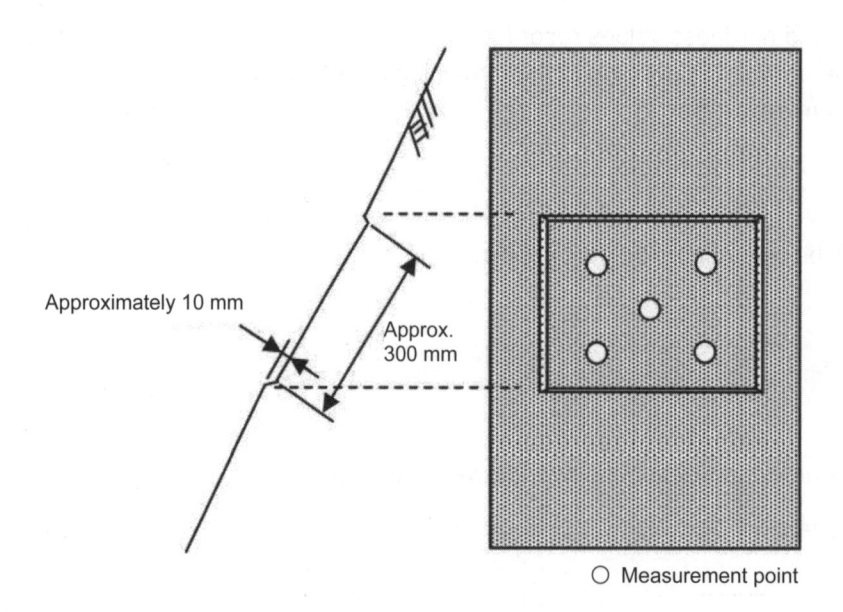

Fig. 2 Shaping of measurement surface and measurement points

Japanese Geotechnical Society Standard (JGS 3411-2012)
Method for rebound hammer test on rocks

1 Scope

This standard specifies methods for determining the hammer rebound value in a rock mass in situ. The hammer rebound value is used for simple estimation of the mechanical properties of a rock mass. This test applies to in situ rock masses from soft rock through hard rock. However, extremely soft rock masses in which localized collapse may occur during the test and prevent suitable measurements from being obtained are excluded from the scope of application.

2 Normative references

None

3 Terms and definitions

The main terms and definitions used in this standard are as follows:

3.1 Schmidt hammer

A Schmidt hammer is a test apparatus in which spring force is used to impact the point to be measured with a hammer in order to measure the rebound quantity.

3.2 Hammer rebound value

The hammer rebound value is the quantity of rebound resulting from impacting.

3.3 Measurement surface

The measurement surface is the exposed surface of the rock mass to be tested. The measurement surface shall be smooth and flat.

3.4 Measurement point

The measurement point is a point on the measurement surface that is established as a point to be impacted.

4 Equipment

4.1 General matters

A calibrated Schmidt hammer shall be used as the test apparatus.

4.2 Plunger shape of Schmidt hammer

Fig. 1 shows an example of the Schmidt hammer. There are two types of Schmidt hammer: one in which the end of the plunger is bar-shaped (Fig. 1 (a)) and a mushroom-shaped one that has a larger impacting surface (Fig. 1 (b)).

Note: For soft rock, the hammer with the mushroom-shaped plunger end should be used.

4.3 Calibration of Schmidt hammer

Before using the Schmidt hammer, it shall be used to impact a metal test anvil whose hammer rebound value is known in order to confirm that the prescribed hammer rebound value can be obtained. If the prescribed hammer rebound value is not obtained, the Schmidt hammer shall be adjusted.

5 Test method

5.1 General matters

In-situ tests shall be conducted by selecting the measurement surface, establishing the measurement points, processing the area around the measurement points and measuring the hammer rebound value in that order.

5.2 Selection of measurement surface

Select a measurement surface that exhibits the representative geological features of the rock mass to be tested (status of rock, discontinuities and so on). As a rule, the measurement surface shall be a square 150 - 500 mm in size.

5.3 Establishment of measurement points

Establish at least nine points within the measurement surface. However, the interval between measurement points shall be equal to or greater than the size of the diameter of the plunger end. If a section of the measurement surface is made up of loosed rocks, do not establish measurement points in that section. Fig. 2 shows an example of the placement of measurement points.

Note: A scan line approximately 1 m in length may be established on the surface of the (planar) rock mass and the measurement points may be established on this line.

5.4 Processing of area around measurement points

If there is surface roughness measuring 1 mm or greater near the measurement points (within the area covered by the end of the plunger), either change the measurement point to a different location or shape the area using a grinder, grindstone or other means so the surface roughness is less than 1 mm. If there is any extraneous material on the measurement surface, remove this material.

5.5 Measurement of hammer rebound value

Hold the body of the Schmidt hammer and place the plunger end against the measurement point. Hold the hammer so the plunger axis is vertical with respect to the measurement surface and press gradually to obtain the hammer rebound value from the action of impact that is produced as the plunger is pressed in. Impact should be applied only once at each measurement point. If readings that can clearly be determined to be mistakes (based on the sound of impact during measurement, the state of the indentation and so on) are obtained, or if the area near the measurement point has collapsed, that measurement value shall be discarded and additional measurements shall be performed to make up for the insufficiency.

6 Test results

For each measurement surface, the average of all readings (2 significant digits) shall be determined and used as the hammer rebound value.

Note: If necessary (for example, when contrasting the results of measurement with different impacting directions), the results shall be corrected with respect to the direction of impact, using a conversion chart or the like.

7 Reporting

The following items shall be reported in the test results.

a) Location of measurement surface

Note 1: If necessary, sketches and photographs of the measurement surface shall be appended to the report.

b) Status of rock mass at measurement surface and surrounding area

Note 2: If necessary, the lithology, type of rock, status of joints, cracks and other discontinuities, status of spring water, rock mass classification and its standard to be adopted, dampness of measurement surface, processing of measurement surface and so on shall be noted.

c) Hammer rebound value

Note 3: If necessary, the model name of the Schmidt hammer shall be noted.

Note 4: If necessary, the individual readings for hammer rebound value, the direction of impact, corrections and so on shall be noted.

JGS 3411-2012

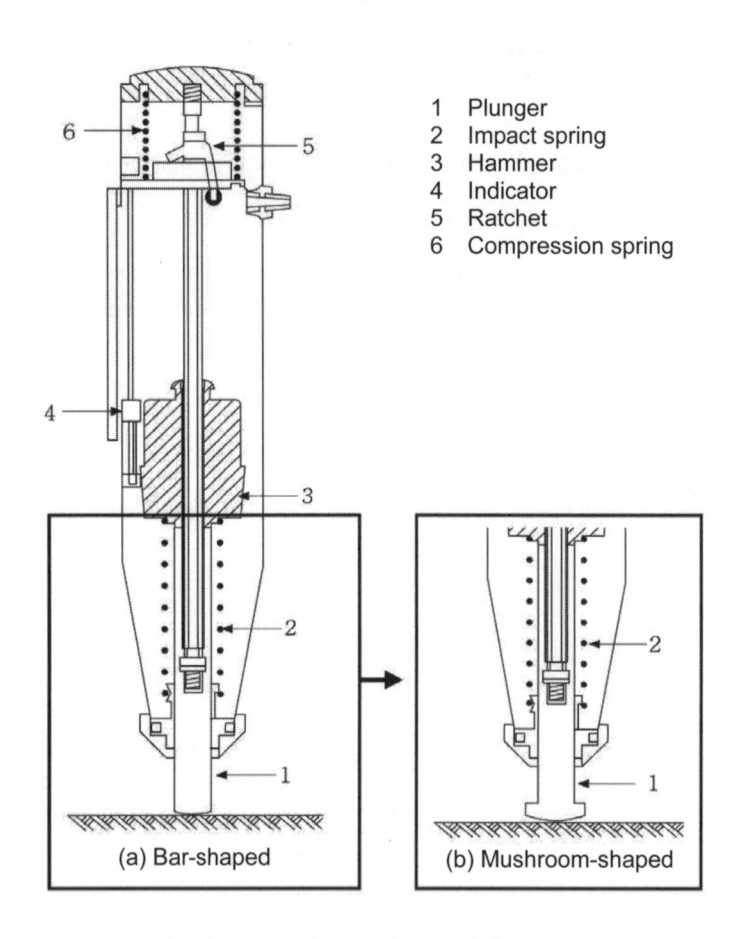

1 Plunger
2 Impact spring
3 Hammer
4 Indicator
5 Ratchet
6 Compression spring

(a) Bar-shaped

(b) Mushroom-shaped

Fig. 1 Examples of Schmidt hammer

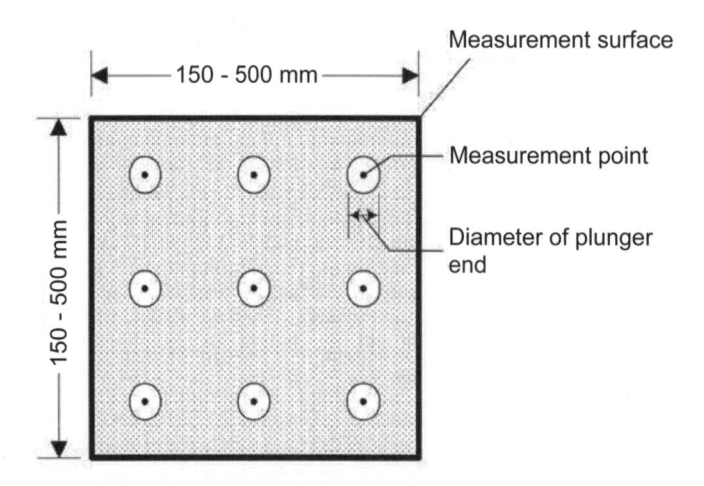

Fig. 2 Example of relationship between measurement surface and measurement points

Japanese Geotechnical Society Standard (JGS 3421-2012)
Method for point load test on rocks

1 Scope

This standard specifies methods for determining the point load strength of rock specimens. Rock specimens comprise rock, gravel and boring cores.

2 Normative references

None

3 Terms and definitions

The main terms and definitions used in this standard are as follows:

3.1 Specimen

Specimens are shaped or unshaped rock used for tests.

3.2 Point load strength (I_s)

Point load strength is the value derived by dividing the fracture load measured in the point load test by the square of the equivalent core diameter.

3.3 Fracture load (P)

The fracture load is the load at which the test specimen fractures.

3.4 Loading point interval (D)

The loading point interval is the distance between two loading points.

3.5 Equivalent core diameter (D_e)

The equivalent core diameter is the diameter of a circle with an area that is equivalent to the cross section of the minimum sectional area of the specimen including loading points.

3.6 Specimen width (W)

The specimen width is the minimum width of the specimen measured in the direction perpendicular to the loading direction.

3.7 Loading point distance (L)

The loading point distance is the distance from the loading axis to the nearest free end face of the specimen, along the direction perpendicular to the loading direction and the specimen width.

3.8 Loading axis

The loading axis is the line connecting the two loading points.

4 Equipment

The test equipment shall comprise a loading section and a load measurement apparatus. The test equipment shall meet the following conditions.

a) The loading section shall comprise a loading frame, a hydraulic jack (or other loading apparatus) and loading cones.

b) In the test equipment, the central axis of the two loading cones shall match the loading axis.

c) The loading cones shall be made of a material with sufficient rigidity.

d) The apex of the loading cones shall have the shape shown in Fig. 1.

e) The loading section shall possess sufficient rigidity and load-carrying capacity with respect to the fracture load of the specimen.

f) The load measurement apparatus shall be capable of measuring the fracture load and shall be able to conduct measurements with an accuracy of ±5% of the fracture load.

5 Test method

5.1 Preparation of specimen

The specimen shall be prepared as follows:

a) From the material that has been sampled from the rock that is to be examined, prepare a specimen of the size that can be tested.

b) Rock, gravel and boring cores shall be used as the specimens. Simple shaping may be performed if necessary to satisfy the following conditions c) and d).

c) The size of the specimen shall be such that the relationship between the loading point interval and the specimen width, and the relationship among the loading point interval, the equivalent core diameter, the specimen width, and the loading point distance shall meet the conditions shown in Fig. 2. The loading point interval shall be $0.3W < D < W$. The loading point distance shall be $L > 0.5D$ in the case of (a), (c), (d) and (e).

d) Confirm the anisotropy of the specimen and the presence of planes of weakness.

e) Measure the specimen width and calculate the equivalent core diameter.

5.2 Loading test

The loading test shall be performed as follows.

a) Prepare the test equipment and confirm the operation at startup.

b) Place the specimen in the test equipment.

c) Measure the loading point interval and the loading point distance. The loading point distance shall satisfy the conditions shown in Fig. 2.

d) Conduct loading at a constant rate of loading until the specimen fractures, and measure the fracture load.

e) Remove the fractured specimen and end the test for that specimen.

f) When continuing the test for other specimens, the test shall be repeated from b) above.

5.3 Recording of test results

The following test results shall be recorded.

a) Specimen number

b) Specimen shape

c) Sampling location

d) Rock name

e) Loading point interval

f) Specimen width

g) Loading point distance

h) Fracture load

i) If the specimen did not fracture at the cross-section of the minimum sectional area, that fact shall be recorded.

j) Other matters of note shall be recorded If necessary.

6 Analysis of test results

The point load strength shall be calculated using the following equation.

$$I_s = \frac{P}{D_e^2}$$

where

I_s: Point load strength (MN/m²)
P: Fracture load (N)
D_e: Equivalent core diameter (mm)
$D_e = D$ when specimen is cylindrical (horizontal)
$D_e^2 = \dfrac{4WD}{\pi}$ in all other cases

7 Reporting

The following items shall be reported.

a) Specimen number

b) Specimen shape (name of specimen shape as shown in Fig. 2)

c) Sampling location

d) Rock name

e) Fracture load

f) Loading point interval

g) Equivalent core diameter

h) Specimen width

i) Loading point distance

j) Point load strength

k) If the method used deviates in any way from this standard, give details of the method used.

l) Other reportable matters

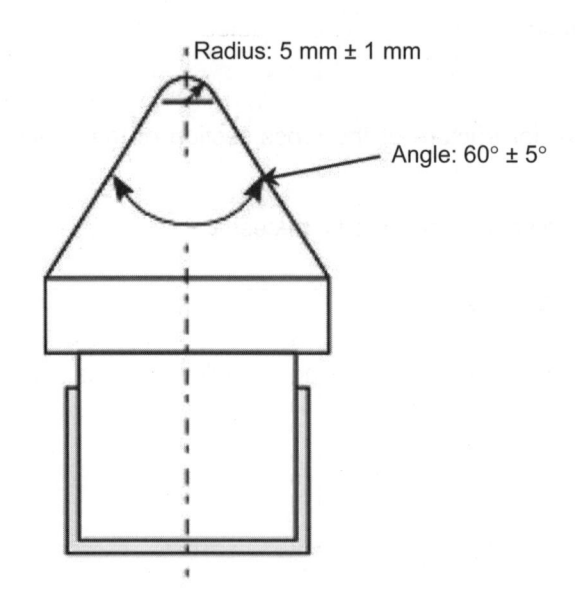

Fig. 1 Shape of loading cone apex

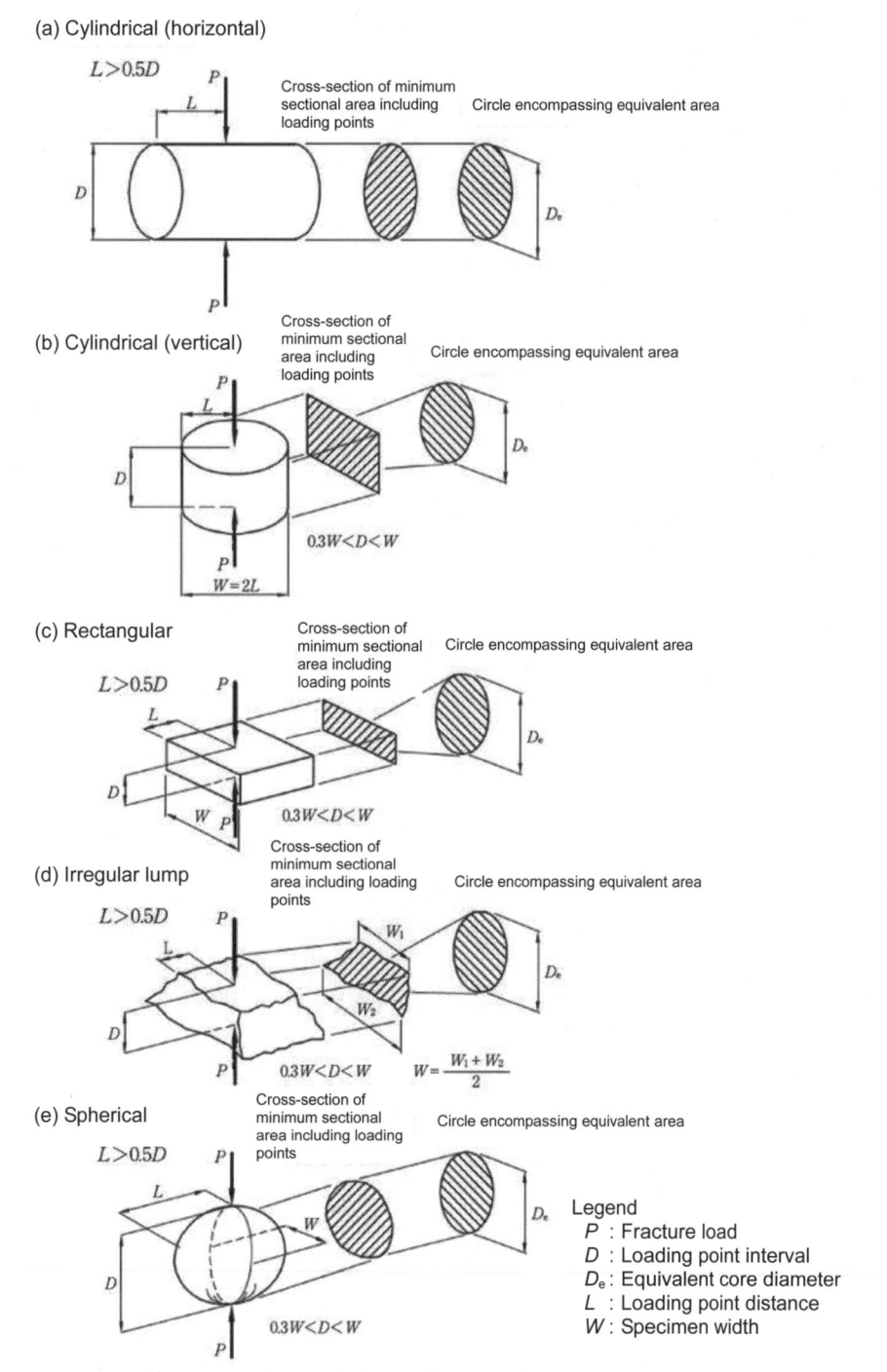

(a) Cylindrical (horizontal)

(b) Cylindrical (vertical)

(c) Rectangular

(d) Irregular lump

(e) Spherical

Legend
P : Fracture load
D : Loading point interval
D_e : Equivalent core diameter
L : Loading point distance
W : Specimen width

Names of specimen shapes: Cylindrical (horizontal), cylindrical (vertical), rectangular, irregular lump, spherical

Fig. 2 Specimen shapes for point load test

Japanese Geotechnical Society Standard (JGS 3431-2012) Method for needle penetration test

1 Scope

This standard specifies methods for penetrating soil or rock with a needle, measuring the penetration length and penetration load, and determining the needle penetration gradient from the result. The test applies primarily to soil and soft rock that can be penetrated by a needle, including solidified soil.

2 Normative references

The following standards constitute a part of this Standard by virtue of being referenced in this Standard. Of these referenced standards, in the case of a standard bearing a Western calendar date, the version for the year noted shall apply, and subsequent revised versions (including supplements) shall not apply. In the case of referenced standards not bearing a Western calendar date, the latest versions of these standards shall apply (including supplements).

JIS S 3008   Hand sewing needles

3 Terms and definitions

The main terms and definitions used in this standard are as follows:

3.1 Needle penetration

Needle penetration means the act of penetration with a needle.

3.2 Needle penetration length (L)

The needle penetration length is the length to which the needle penetrates the object of measurement.

Note: The needle penetration length shall be expressed in mm.

3.3 Penetration load (P)

The penetration load is the resisting force with respect to needle penetration. It is also referred to as the penetration force.

Note: The penetration load shall be expressed in N.

3.4 Needle penetration gradient (N_p)

The needle penetration gradient is the value of the penetration load (P) divided by the penetration length (L).

4 Equipment

The types of test apparatus used for the test comprise a portable type and a desktop type. The test apparatus shall fulfill the following requirements.

4.1 Configuration of portable needle penetration test apparatus

The portable needle penetration test apparatus shall have the configuration shown in Fig. 1.

4.1.1 Configuration

The apparatus shall comprise a needle, a chuck, a needle penetration length measurement section, and a spring compression measurement section that converts this value into a penetration load value.

4.1.2 Penetration needle

The penetration needle noted in JIS S 3008 as "cotton needle No. 2, large blind stitching needle (ø0.84 or 0.89 mm, length (54.5±1.4) mm" or the equivalent shall be used in the section of soil or rock to be penetrated. The needle shall not be broken, bent or worn.

Note: The standard needle shall be the "cotton needle No. 2, large blind stitching needle" noted in JIS S 3008. Application of the equivalent needle shall be allowed based on a consideration of toughness, wear resistance, degree of danger in the event of breakage and so on.

4.1.3 Chuck

The aforementioned needle shall be mounted on the chuck and it shall be possible to fix the projection length at (10±0.2) mm. The chuck shall be able to properly transmit the penetration load.

4.1.4 Needle penetration length measurement section

This section shall be able to measure the needle penetration length accurately at least down to the mm unit.

4.1.5 Spring

The spring shall be a coil spring mounted inside, connected to the needle chuck section, and shall be contracted in proportion with the load during needle penetration.

4.1.6 Penetration load indicator section

This section shall indicate the penetration load from the amount of spring compression. Alternately, it shall have a scale with the compression level converted into a penetration load value. A floating indicator or the like shall be used to indicate the value.

4.2 Desktop needle penetration test apparatus

The desktop needle penetration test apparatus shall have the configuration shown in Fig. 2.

4.2.1 Penetration needle

The penetration needle noted in JIS S 3008 as "cotton needle No. 2, large blind stitching needle (ø0.84 or 0.89 mm, length (54.5±1.4) mm" or the equivalent shall be used in the section of soil or rock to be penetrated. The needle shall not be broken, bent or worn.

Note: The standard needle shall be the "cotton needle No. 2, large blind stitching needle" noted in JIS S 3008. Application of the equivalent needle shall be allowed based on a consideration of toughness, wear resistance, degree of danger in the event of breakage and so on.

4.2.2 Chuck

The aforementioned needle shall be mounted on the chuck and it shall be possible to fix the projection length at (10±0.2) mm. The chuck shall be able to properly transmit the penetration load.

4.2.3 Displacement gauge

The displacement gauge shall be able to accurately measure the amount of vertical displacement that corresponds to the needle penetration length, at least down to the mm unit.

4.2.4 Elevator stage

The elevator stage shall be a mechanism that moves the sample or the needle, up or down for the purpose of needle penetration.

Note: If there is no elevator stage, the compression frame shall be movable, and the displacement gauge shall also measure the displacement on the compression frame side.

4.2.5 Penetration load measurement section

This section shall measure the penetration load using a load cell or a force gauge.

5 Test method

The test method shall be as follows.

5.1 Portable test apparatus

When the portable test apparatus is used, the test procedure shall comprise in order, the following steps; preparation, needle penetration, and readout of penetration length and penetration load.

5.1.1 Preparation

Check the status of the needle to ensure that it is not broken or bent, and that the tip is not worn. Then set the floating indicators, etc., for the needle penetration length measurement section, the penetration load measurement section, and so on, to the zero position.

5.1.2 Needle penetration

Place the tip of the needle vertically against the measurement surface and penetrate gradually at a constant rate of speed ensuring that there is no eccentricity.

5.1.3 Readout of penetration length and penetration force

When the needle penetration length reaches 10 mm, or when the penetration load (P) reaches the maximum value of the test apparatus, read the values for the needle penetration length (L) and the penetration load (P).

5.2 Desktop test apparatus

The test procedure when the desktop test apparatus is used shall be as follows.

5.2.1 Preparation

Check the status of the needle to ensure that it is not broken or bent, and that the tip is not worn. Then read the initial values for the needle penetration length measurement section and the penetration load measurement section.

5.2.2 Needle penetration

Place the tip of the needle vertically against the measurement surface and penetrate gradually at a constant rate of speed ensuring that there is no eccentricity. The standard needle penetration speed shall be (20±5) mm/min.

5.2.3 Readout of needle penetration length and penetration load

When the needle penetration length reaches 10 mm, or when the penetration load (P) reaches the maximum value of the test apparatus, read the values for the needle penetration length (L) and the penetration load (P). If it is possible to read the penetration load for each 1 mm of needle penetration length, record these values (Fig. 3).

6 Analysis of results

Calculate the needle penetration gradient (N_p).

6.1 Calculation of needle penetration gradient

Calculate the needle penetration gradient (N_p) from the relationship between the needle penetration length (L) and the penetration load (P), using the following equation.

$$N_P = \frac{P}{L}$$

where

N_P : Needle penetration gradient (N/mm)
P: Penetration load (N)
L: Needle penetration length (mm)

7 Reporting

The following items shall be reported as test results.

a) Purpose of test

b) Specimen (measurement point) number

c) Test location

d) Specimen preparation method

e) Needle penetration length L (mm)

f) Penetration load P (N)

g) Needle penetration gradient N_P (N/mm)

i) Other reportable matters

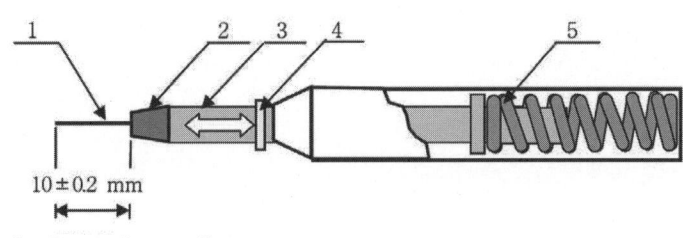

1 Penetration needle
2 Chuck
3 Spindle
4 Floating indicator
5 Spring

Fig. 1 Portable needle penetration test apparatus

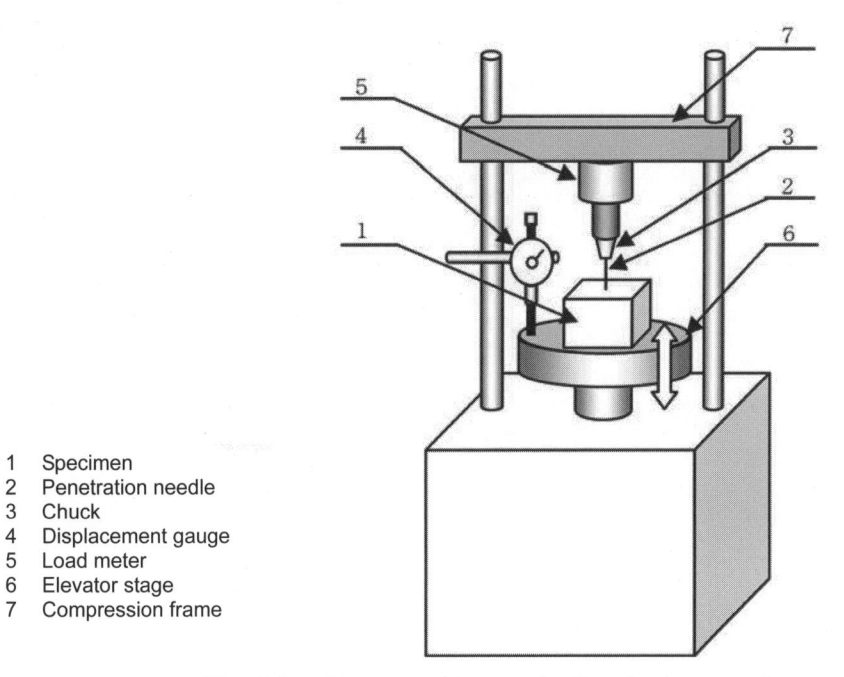

1 Specimen
2 Penetration needle
3 Chuck
4 Displacement gauge
5 Load meter
6 Elevator stage
7 Compression frame

Fig. 2 Desktop needle penetration test apparatus

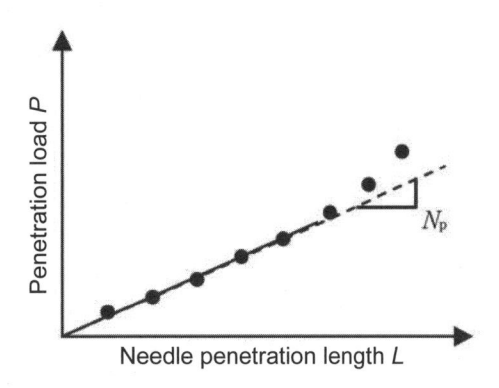

Fig. 3 Method for determining needle penetration gradient (desktop type)

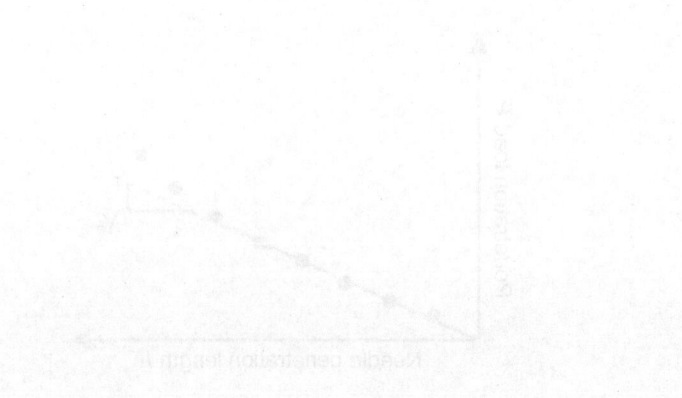

Japanese Geotechnical Society Standard (JGS 1611-2012)
Test method for soil density by the compacted sand replacement method

1 Scope

This standard specifies methods for testing the density of in-situ soil using the compacted sand replacement method. The range of soil that can be tested using the apparatus and method specified in this standard is limited to soil with a maximum grain size of 150 mm.

2 Normative references

The following standards shall constitute a part of this standard by virtue of being referenced in this standard. The latest versions of these standards shall apply (including supplements).

JIS A 1101   Method of test for slump of concrete
JIS A 1203   Test method for water content of soils
JIS A 1214   Test method for soil density by the sand replacement method
JIS Z 8801-1 Test sieves - Part 1: Test sieves of metal wire cloth

3 Terms and definitions

The main terms and definitions used in this standard are as follows:

3.1 Compacted sand replacement method

A method for determining volume by replacing the soil in the testing hole with sand after confirming to get constant density by the specified compaction method.

4 Types of test method and their selection

4.1 Types of test method

The test methods shall be the three test methods shown in Table 1.

4.2 Selection of test method

The test method shall be selected from the methods shown in Table 1 in accordance with the maximum grain size of the soil particles or rock particles.

Table 1 shows the names of the test methods and the corresponding maximum grain size, the general dimensions, and the number of blows.

Note: The maximum grain size should be determined by using a sieve or conducting a visual inspection in the sampled soil.

5 Equipment

5.1 Baseplate and upper frame (collar)

The baseplate and upper frame (collar) shall satisfy the following conditions.

Note: Fig. 1 shows an example of a baseplate and upper frame (collar). The central projections are not absolutely necessary, but they should be used for smoothing the surface of the sand horizontally.

a) Baseplate

The baseplate shall be a metal plate with a hole with a diameter of (150±0.6) mm, (250±1.0) mm or (300±1.0) mm in the center.

b) Upper frame (collar)

The upper frame (collar) shall have an inner diameter of (150±0.6) mm, (250±1.0) mm or (300±1.0) mm and a height of (50±0.5) mm, and shall be able to be mounted on the baseplate.

5.2 Compacting rod

The compacting rod shall be a round steel or metal rod with diameter (16±0.2) mm and length 500 - 600 mm as specified in JIS A 1101, and shall be semi-spherical at an end.

5.3 Calibration container

The calibration container shall be a container with inner diameter (150±0.6) mm and depth (150±0.6) mm, inner diameter (250±1.0) mm and depth (200±0.8) mm, or inner diameter (300±1.0) mm and depth (300±1.0) mm.

Note: Fig. 2 shows an example of a calibration container.

5.4 Test sand

The test sand shall be sand with a grain size distribution that is able to pass through the 2 mm opening but remains in the 75 μm opening of the metal wire mesh sieve specified in JIS Z 8801-1, and shall be rinsed and dried thoroughly.

Note: The test sand shall be dried thoroughly in an oven. An artificial material (glass beads, etc.) that meets the specified grain size range and has stable properties may be used instead of test sand.

5.5 Scale

The scale shall be able to display readout values up to 1 g in the case of test method A or B and up to 10 g in the case of test method C.

5.6 Other equipment

Other equipment shall be as follows.

a) Sieve

The sieve shall be the metal wire mesh sieve specified in JIS Z 8801-1.

Note: The rectangular perforated metal plate sieve specified in JIS Z 8801-2 may also be used.

b) Water content measurement apparatus

The water content measurement apparatus shall be the one specified in JIS A 1203.

c) Straight knife

The straight knife shall be a metal knife with a single blade measuring 300 mm or more in length. It shall have a length that will not impede operations.

d) Test hole excavation apparatus

The test hole excavation apparatus shall be the one specified in JIS A 1214.

e) Excavated soil storage bag

The excavated soil storage gear shall be a plastic bag or container.

6 Test method

6.1 Calibration of test sand density

The density of the test sand shall be calibrated as follows.

a) Weigh the mass m_1 (g) of the calibration container.

b) Place the upper frame (collar) on the top of the calibration container and, from the height of the upper frame (collar), gently pour test sand into the container up to the upper surface of the upper frame (collar). Smooth the surface of the sand gently to the same level of the height of the upper surface of the upper frame (collar).

c) Penetrate the compacting rod by hand so that the sand is spread uniformly throughout the entire cross-sectional area and compact the sand for the number indicated in Table 1 to be uniformly densified.

The compacting rod shall be penetrated until its end reaches to approximately 80% of the depth of the container. Fig. 3 schematically shows the position of the calibration container and compacting rod.

d) Remove the upper frame (collar) and smooth the sand at the upper surface of the test container by using the straight knife. Weigh the mass m_2 (g) of the calibration container and sand.

e) Conduct the procedure of b) - d) at least three times. If the difference between the maximum and minimum values for $m_2 - m_1$ (g) derived from the measurements is within 0.85% of the average value, it shall be used as the result of calibration of the measurements. If the difference between the maximum and minimum values exceeds 0.85%, repeat the procedure of b) - d) until the difference satisfies the above figure.

Note: The density of the test sand will vary depending on the water content and grain size of the test sand, and this will affect the test results. For this reason, when the test sand is transported and stored and when conducting site testing, proper management of test sand should be conducted to ensure that the values do not change from the water content and grain size during calibration.

6.2 Measurement of soil density

The soil density shall be measured as follows.

a) Using the straight knife, smooth the surface of the ground at the test location. During this process, if the surface of the ground contains loosened soil, rocks or rubbish, remove these items.

b) Place the baseplate on the ground surface that has been smoothed to be level, and make the baseplate be tightly in contact with the ground.

c) Excavate the soil on the inside of the baseplate hole using the test hole excavation apparatus. Make sure the hole is as vertical as possible and the surrounding soil is not disturbed. Put all of the excavated soil in the excavated soil storage bag and make sure the water content does not change.

Note: The depth of excavated hole should be measured by a scale or steel tape.

d) Weigh the total mass m_3 (g) of the wet soil that has been removed from the test hole.

e) Thoroughly mix the entire quantity of wet soil whose mass was weighed in d), and measure the water content w (%) of the wet soil by using the method specified in JIS A 1203.

f) Put the test sand in a plastic bag or container and weigh its mass m_4 (g).

g) Place the upper frame (collar) on the top of the baseplate and pour the test sand into the hole using the same method specified in 6.1 b). Then penetrate the compacting rod by hand until the end of the rod is at approximately 80% of the depth of the hole. Compact uniformly the sand for the number indicated in Table 1.

h) Remove the upper frame (collar) and smooth the surface of the sand along the upper surface of the baseplate using the straight knife. Collect the test sand remaining on the top of the baseplate in the plastic bag or container described in 6.2 f) and measure its mass m_5 (g).

7 Test results

Processing of test results shall be carried out as follows.

7.1 Test sand density

Calculate the density of the test sand using equation (1) below.

$$\rho_{ds} = \frac{m_2 - m_1}{V} \tag{1}$$

where

ρ_{ds} : Test sand density (g/cm³)
m_2 : Mass of calibration container and sand (g)
m_1 : Mass of calibration container (g)
V : Volume of calibration container
　　V of container with inner diameter 150 mm = 2,651 cm³
　　V of container with inner diameter 250 mm = 9,817 cm³
　　V of container with inner diameter 300 mm = 21,206 cm³

7.2 Soil density

Calculate the density of the soil as indicated below.

a) Calculate the volume of the test hole using equation (2) below.

$$V_0 = \frac{m_4 - m_5 - m_p}{\rho_{ds}} \tag{2}$$

where

V_0 : Volume of test hole (cm³)
m_4 : Mass of test sand and plastic bag or container (g)
m_5 : Mass of remaining test sand and plastic bag or container (g)
m_p : Mass of sand in baseplate thickness section (g)
　　(hole area x baseplate thickness x ρ_{ds})

b) Calculate the wet density of the soil using equation (3) below.

$$\rho_t = \frac{m_3}{V_0} \tag{3}$$

where

ρ_t : Wet density of sand (g/cm³)
m_3 : Total mass of wet soil extracted from test hole (g)

c) Calculate the dry density of the soil using equation (4) below.

$$\rho_d = \frac{\rho_t}{1 + (w/100)} \tag{4}$$

where

ρ_d : Dry density of sand (g/cm^3)

w : Water content (%)

8 Reporting

The following items shall be reported as test results.

a) Site number and location

b) Test date

c) Test personnel

d) Test method (report the name shown in Table 1)

e) Maximum grain size (mm)

Note: In the case of soil containing gravel and rocks, report the maximum grain size in the soil extracted from the test hole.

f) Water content (%)

g) Wet density (g/cm^3)

h) Dry density (g/cm^3)

i) If the method used deviates in any way from this standard, give details of the method used.

j) Other reportable matters

Table 1 Test method

Test name	Maximum grain size (mm)	Hole diameter (mm)	Hole depth (mm)	No. of compactions
A	53	150	150	15
B	100	250	200	35
C	150	300	300	50

Note a): The baseplate thickness is the thickness of the baseplate in the hole section.

Fig. 1 Example of baseplate and upper frame (collar)

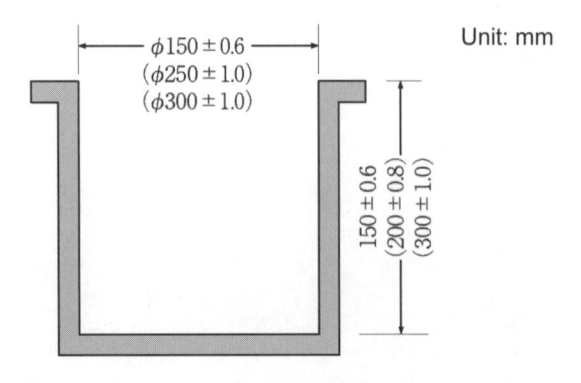

Fig. 2 Example of calibration container

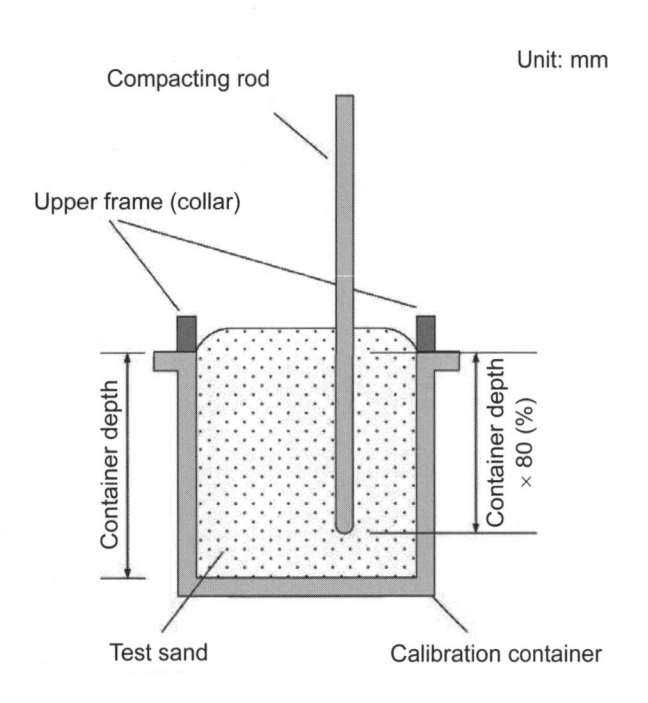

Fig. 3 Relative location of calibration container and compacting rod

Japanese Geotechnical Society Standard (JGS 1612-2012)
Test method for soil density by the water replacement method

1 Scope

This standard specifies methods for determining soil density by excavating a test hole and spreading a sheet so it adheres tightly to the hole wall, and then pouring water into the hole and measuring the volume and determining the soil density based on this value and the quantity of excavated soil. This standard specifies the method for in-situ soil density tests for rock materials and other situations in which it is difficult to measure the density using the sand replacement method and the like.

2 Normative references

The following standards shall constitute a part of this standard by virtue of being referenced in this standard. The latest versions of these standards shall apply (including supplements).

JIS A 1203   Test method for water content of soils
JIS Z 8801-1   Test sieves - Part 1: Test sieves of metal wire cloth
JGS 0051   Method of classification of geomaterials for engineering purposes
JGS 0132   Test method for particle size distribution of rock

3 Terms and definitions

The main terms and definitions used in this standard are as follows:

3.1 Water replacement method

A method for determining the volume of a test hole that has been excavated based on the volume of water needed when water is poured into the test hole.

3.2 Maximum grain size

The grain size indicated by the smallest opening of a sieve through which all of a sample will pass.

Note: In the case of a sample that includes grain sizes of 300 mm or greater, the size of the rock particles may be measured to determine the maximum grain size. For the size of rock particles, the length of the axis with the largest outer diameter is the large diameter; the length of the axis that is perpendicular to this axis and has the largest outer diameter is the medium diameter, and the length of the axis that is perpendicular to both the large diameter and the medium diameter and has the largest outer diameter is the small diameter. As a rule, the maximum grain size in this situation is the medium diameter.

4 Equipment

4.1 Baseplate

The baseplate shall be a metal plate or wooden board with a collar that has a hole with a diameter at least three times the maximum grain size of the material to be measured.

Note: The baseplate hole need not be circular, but in many cases the hole is made round to ensure that the sheet adheres tightly to the walls of the test hole. The collar height shall be approximately 5 - 15% of the hole diameter. Fig. 1 shows an example of a baseplate.

4.2 Sheet

The sheet shall be a clear sheet that is flexible to enable it to adhere tightly to the walls of the test hole.

Note: The sheet should be the vinyl film specified in JIS K 6732 "Poly (vinyl chloride) films for agriculture" with a thickness of approximately 0.15 - 0.3 mm, in order to prevent breakage and minimize measurement error.

4.3 Test water

The test water shall be tap water, well water or other fresh water.

Note: River water may be used if it does not contain refuse, soil particles or other impurities that will affect measurement results.

4.4 Water tank

The water tank shall be one that enables water to be poured gently into the test hole and will enable the total mass to be measured before and after the water is poured in.

Note: The tank capacity should be appropriate for the volume of the excavated section and shall become full in 1 - 5 fill processes. If a flowmeter is used, one with a minimum scale increment of 0.001 m^3 or less shall be used.

4.5 Sieve

The sieve shall be as follows.

a) Sieve: The sieve shall be the metal wire mesh sieve specified in JIS Z 8801-1 with openings of 75 mm and 125 mm.

Note: A rectangular perforated metal plate sieve as specified in JIS Z 8801-2 "Test sieves-. Part 2: Test sieves of perforated metal plate" that has equivalent performance to that of the metal wire mesh sieve may also be used.

b) Single-opening sieve: The single-opening sieve shall have a nominal size of 300 mm.

Note 1: Fig. 2 shows an example of a single-opening sieve. However, the tolerance shall be (±3)% of the sieve opening.

Note 2: If necessary, a single-opening sieve with a sieve opening of other than 300 mm shall be prepared.

4.6 Other equipment

Other equipment shall be as follows.

a) Water content measurement apparatus: The water content measurement apparatus shall be the one specified in JIS A 1203.

b) Test hole excavation gear: The apparatus used to excavate the test hole shall be as follows.

    1) Pick, pickaxe or similar tool

    2) Scoop, etc.

Note: An excavating machine that can excavate without disturbing the hole wall may be used.

c) Scale: The scale shall be capable of displaying the mass of excavated soil down to 1/2000 or lower.

d) Thermometer: The thermometer shall has a minimum scale increment of 1 °C or lower.

e) Level

f) Sand: Sand used to shape the ground surface.

Note: The sand to be used should be clean sand with a grain size of 2 mm or less.

g) Excavated soil storage gear: The excavated soil storage gear shall be a plastic bag or container.

Note: When the excavated soil is stored temporarily, a waterproof sheet or the like shall be used to ensure that the water content. etc. does not change.

5 Test method

5.1 Shaping of test location

The test location shall be shaped as follows.

a) Shape the surface of the ground at the test location so it is as level as possible. During this process, remove any loose soil, rocks, refuse and the like from the ground surface and use sand to smooth any unevenness in the section where the baseplate is placed.

b) Using a level, measure the levelness of the ground surface. The ground surface shall be shaped so that the levelness is within the range indicated by equation (1) as follows. Fig. 3 shows the status of the test after shaping.

$$i < \frac{H}{2D} \tag{1}$$

where

i : Levelness of ground surface
H : Collar height (mm)
D : Test hole diameter (mm)

Note: Determine the levelness i of the ground surface using equation (2) as follows. Fig. 4 shows an example of the measurement method.

$$i = \frac{h}{l} \tag{2}$$

where

h : Placement height of batten
l : Length of batten

5.2 Measurement of volume of baseplate collar section

Measure the volume of the baseplate collar section as follows.

a) Fasten the baseplate on the surface of the ground that has been shaped.

b) Fasten a sheet tightly along the inner wall of the collar and the surface of the ground.

c) Gently fill the hole with water from the water tank up to almost the level at which the water overflows from the top of the collar.

Note: Allowing one section of the sheet to sag as the water is poured in will allow the air to escape from the space between the sheet and the hole wall and improve adhesion. Fig. 5 shows the measurement of the volume of the baseplate collar section.

d) Weigh the mass of the water tank and water before water is poured in m_1 (kg) and the mass of the water tank and water after water is poured in m_2 (kg).

e) Dump the water contained in the sheet at a location where this will not affect the test and then carefully remove the sheet.

5.3 Excavation of test hole and measurement of excavated soil mass

Excavation of the test hole and measurement of the mass of excavated soil shall be done as follows.

a) Using the excavation gear, excavate the test hole along the baseplate. The wall of the test hole shall be smoothed to enable the sheet to adhere tightly.

b) The depth of the test hole shall be 0.6 - 0.7 times the test hole diameter.

c) Use a single-opening sieve, etc. to separate the soil material and the rocks in the excavated soil at the site. Weigh the wet mass of the soil material m_{t1} (kg) and the wet mass of the rocks m_{t2} (kg).

d) Place the excavated soil in the excavated soil storage gear to protect it and ensure that the water content does not change. Shape the ground surface at the test location so it is as smooth as possible. During this process, remove any loose soil, rocks, refuse and the like from the ground surface and use sand to smooth any unevenness in the section where the baseplate is placed.

5.4 Measurement of volume of test hole

The volume of the test hole shall be measured as follows.

a) Spread the sheet so it will adhere easily to the wall of the test hole.

b) Fill the hole with water from the water tank up to the top of the collar.

c) Measure the mass m_3 (kg) of the water tank and water before the hole is filled with water and the mass m_4 (kg) of the water tank and water after the hole is filled with water.

d) Measure the temperature t (°C) of the water that is used.

Note: If the water tank is small and the mass m_3 (kg) and m_4 (kg) is measured in several increments, the value shall be the cumulative value of these measurements. Fig. 6 shows the method used to measure the volume of the excavation section.

5.5 Measurement of water content of excavated soil

The water content of the excavated soil shall be measured by determining the individual water content and then determining the weighted average using the dry mass ratio as shown in 6.1.

a) Determine the water content of the soil material w_f (%) as described in JIS A 1203.

b) If measurement of the water content of all soil material is difficult due to the configuration of the drying oven or the like, the soil material shall be stirred thoroughly so the overall water content will be the same, and then soil material samples that have been sampled appropriately shall be subjected to water content testing.

c) Determine the water content of the rocks w_c (%) as described in JIS A 1203, etc.

Note: In addition to the method in JIS A 1203, the water content of the rocks w_c (%) is sometimes determined by using the coefficient of water absorption as shown in JGS 2134 "Method for moisture content test on rocks" and JIS A 1110 "Methods of test for density and water absorption of coarse aggregates" as the water content.

d) A water content test for the rocks shall be conducted by means of an extraction test conducted by appropriate sampling in accordance with the typical grain size, type of rock and degree of weathering.

Note: If the material includes sand, gravel, weathered rock and other different types of materials, the samples shall be selected taking into consideration the frequency of occurrence of different materials.

6 Test results

6.1 Dry mass of soil material

Calculate the dry mass of the soil material using equation (3) below.

$$m_{d1} = \frac{m_{t1}}{1+(w_f/100)} \tag{3}$$

where

m_{d1}: Dry mass of soil material (kg)
w_f: Water content of soil material (%)
m_{t1}: Wet mass of soil material (kg)

6.2 Dry mass of rocks

Calculate the dry mass of the rocks using equation (4) below.

$$m_{d2} = \frac{m_{t2}}{1+(w_c/100)} \tag{4}$$

where

m_{d2}: Dry mass of rocks (kg)
w_c: Water content of rocks (%)
m_{t2}: Wet mass of rocks (kg)

6.3 Dry mass ratio of soil material

Calculate the dry mass ratio of the soil material using equation (5) below.

$$P_f = \frac{m_{d1}}{m_{d1}+m_{d2}} \tag{5}$$

where

P_f: Ratio of dry mass of soil material as compared to total quantity of excavated soil

6.4 Water content of excavated soil

Calculate the water content of all excavated soil using equation (6) below.

$$w = w_f P_f + w_c(1-P_f) \tag{6}$$

where

w: Water content of all excavated soil (%)

6.5 Volume of soil excavated from test hole

Calculate the volume of soil excavated from the test hole using equation (7).

$$V = V_2 - V_1 = \frac{1}{1000\rho_w}(m_3 - m_4) - \frac{1}{1000\rho_w}(m_1 - m_2) \tag{7}$$

where

- V : Volume of soil excavated from test hole (m³)
- V_1 : Volume of baseplate section (m³)
- V_2 : Volume of baseplate section and test hole (m³)
- m_1 : Mass of water tank before pouring water into baseplate collar section (kg)
- m_2 : Mass of water tank after pouring water into baseplate collar section (kg)
- m_3 : Mass of water tank before pouring water into test hole (kg)
- m_4 : Mass of water tank after pouring water into test hole (kg)
- ρ_w : Density of water (g/cm³) at water temperature t (°C) (see Table 1)

6.6 Wet density and dry density of soil

Determine the wet density and dry density of the soil using the following procedure.

a) Calculate the wet density of the soil using equation (8).

$$\rho_t = \frac{m_{t1} + m_{t2}}{1000V} \tag{8}$$

where

ρ_t : Wet density of soil (g/cm³)

b) Calculate the dry density of the soil using equation (9).

$$\rho_d = \frac{\rho_t}{1 + (w/100)} \tag{9}$$

where

ρ_d : Dry density of soil (g/cm³)

7 Reporting

The following items shall be reported as test results.

a) Site number and location
b) Test date
c) Test personnel
d) Maximum grain size (mm)
e) Water content (%)
f) Wet density (g/cm³)
g) Dry density (g/cm³)
h) Test hole diameter, depth and status
i) If the method used deviates in any way from this standard, give details of the method used.
j) Other reportable matters

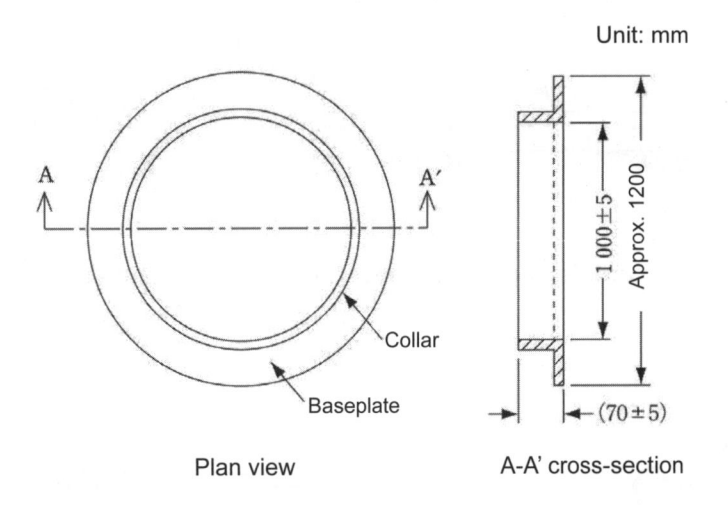

Fig. 1 Example of baseplate with collar

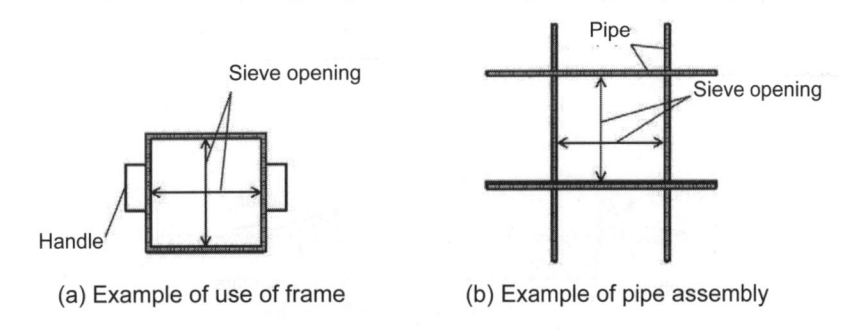

Fig. 2 Example of single-opening sieve

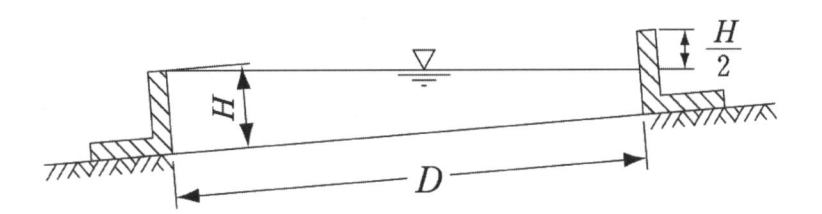

(a) Example of use of frame

(b) Example of pipe assembly

Fig. 3 Status of test after shaping

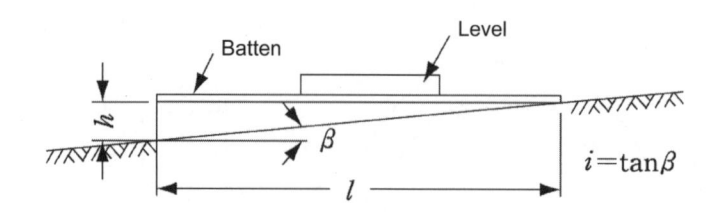

Fig. 4 Example of gradient measurement of ground surface

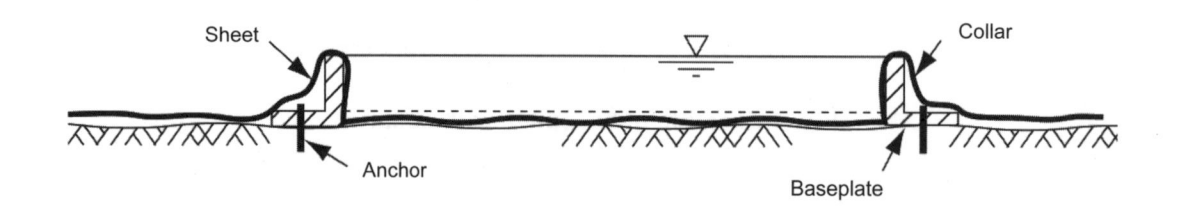

Fig. 5 Measurement of volume of baseplate collar section

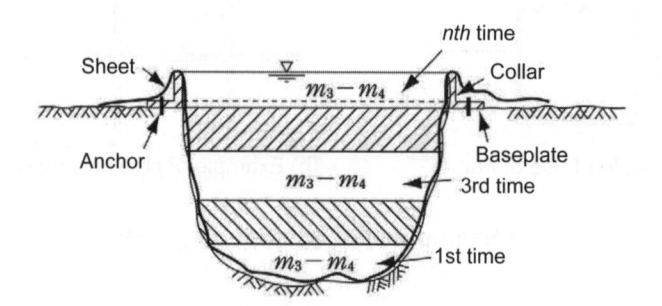

Fig. 6 Measurement of volume of excavated section

Table 1 Water density

Water temperature t (°C)	Water density ρ_w (g/cm^3)	Water temperature t (°C)	Water density ρ_w (g/cm^3)	Water temperature t (°C)	Water density ρ_w (g/cm^3)
4	1.000 0	16	0.998 9	28	0.996 2
5	1.000 0	17	0.998 8	29	0.995 9
6	0.999 9	18	0.998 6	30	0.995 7
7	0.999 9	19	0.998 4	31	0.995 3
8	0.999 9	20	0.998 2	32	0.995 0
9	0.999 8	21	0.998 0	33	0.994 7
10	0.999 7	22	0.997 8	34	0.994 4
11	0.999 6	23	0.997 5	35	0.994 0
12	0.999 5	24	0.997 3	36	0.993 7
13	0.999 4	25	0.997 0	37	0.993 3
14	0.999 2	26	0.996 8	38	0.993 0
15	0.999 1	27	0.996 5	39	0.992 6

Japanese Geotechnical Society Standard (JGS 1613-2012)
Test method for soil density using core cutter

1 Scope

This standard specifies methods for testing the density of in-situ soil using a core cotter. The range in which tests can be conducted using the equipment and methods specified in this standard shall be soil in which there is no barrier to penetration by a core cutter and from which suitable results can be obtained.

2 Normative references

The following standard shall constitute a part of this standard by virtue of being referenced herein. The latest edition of referenced standard shall apply (including supplements).

● JIS A 1203   Test method for water content of soils

3 Terms and definitions

The main terms and definitions used in this standard are as follows:

3.1 Core cutter

A steel cylinder with a known inner cavity whose lower edge has a blade.

4 Equipment

4.1 Core cutter

The core cutter shall be a steel cylinder with an inner diameter of 50 - 150 mm and a wall thickness of 1 - 4 mm, with a lower edge having a blade for easy penetration.

A core cutter with a cutting edge that is significantly deformed shall be replaced, or the blade shall be sharpened. A core cutter with a height that is 0.8 - 1.3 times the inner diameter shall be selected in accordance with the thickness of the layer to be measured.

Note 1: Fig. 1 shows an example of a core cutter.

The cross-sectional area C_a (%) of the core cutter shall be calculated using equation (1) below.

$$C_a = \frac{D_e^2 - D_i^2}{D_i^2} \times 100 \qquad (1)$$

where

D_e : Maximum outer diameter of core cutter (mm)
D_i : Minimum inner diameter of core cutter (mm)

A core cutter with a C_a value of approximately 10% should be used.

Note 2: The interior of the core cutter should be finished so it is smooth in order to reduce the friction between the core cutter and the soil during penetration.

Note 3: The cutting edge of the core cutter is easily damaged during penetration, so a spare should be prepared prior to conducting measurements at the site.

Note 4: An existing mold may be used in the same manner as a core cutter if an apparatus that can be attached to provide the effectiveness of a cutting edge is used, such as the cutter ring specified in JIS A 1211.

Note 5: A thin wall sampling tube may be processed and used in place of a core cutter. Fig. 2 shows an example in which a thin wall sampling tube is used. The height should be 100 or 150 mm, the inner diameter should be 50 or 75 mm, the wall thickness should be 1.5 - 2.0 mm, the angle of the cutting edge should be (6±1)°, and the thickness of the cutting edge should be (0.2±0.05) mm.

4.2 Core cutter head

The core cutter head shall be attached to the top of the core cutter to form an apparatus that can be used to press the core cutter into the soil or drive it into the soil using a hammer.

Note: Fig. 3 shows an example of the coupled core cutter and core cutter head.

4.3 Hammer

The hammer shall be a mallet or the like that is used to apply the proper impacts to the core cutter head to cause the core cutter to penetrate the soil.

Note: If an apparatus that can apply static force to the core cutter head in order to cause the core cutter to penetrate the soil is available, it is desirable that such apparatus be used.

4.4 Scale

The balance shall be able to indicate readings in increments of 0.01 g or less. The balance shall be selected in accordance with the size of the core cutter.

4.5 Other equipment

Other equipment shall be as follows.

a) Water content measurement apparatus

   The water content measurement apparatus shall be the one specified in JIS A 1203.

b) Excavation tools

   The excavation tools shall comprise a hand scoop, spoon, spatula, brush and so on.

c) Straight knife

   The straight knife shall be a steel knife with a single blade, measuring 250 mm or more in length.

d) Vernier caliper

   The vernier caliper shall have a minimum scale increment of 0.1 mm or less.

e) Sample extruder

   The sample extruder shall be able to push out the soil inside the core cutter.

Note: The sample extruder shall be a jack or similar apparatus. However, if it is difficult to push out the soil using the sample extruder, the soil may be removed from the core cutter using a spatula or trowel.

5 Test method

5.1 Measurement of core cutter mass and volume

The mass and volume of the core cutter shall be measured as follows.

a) Use a balance to measure the core cutter mass m_1 (g).

Note: When measuring the core cutter mass, check to make sure there is no mud or the like sticking to the core cutter.

b) Determine the volume V (cm^3) of the core cutter. To determine the volume, use a vernier caliper and measure the height H (mm) of the core cutter and the inner diameter D (mm) of the cutting edge section to an accuracy of 0.1 mm.

To determine the inner diameter, use the average of diameter measurements taken at right angles to one another.

5.2 Measurement of soil density

The soil density shall be measured as follows.

a) Smooth the surface of the ground at the test location so it is level. During this process, remove any loose soil, rocks or rubbish on the surface of the ground.

Note: If the surface to be measured is not the surface of the ground, use a scoop or the like to excavate down to the required depth, providing an additional margin of width to facilitate measurement operations, and then smooth the surface so it is level. During this process, remove any loose soil, rocks or rubbish on the surface of the ground.

b) Attach the core cutter to the core cutter head and place the blade face down against the level measurement surface. Then cause it to penetrate the soil, taking care to ensure that it oscillates as little as possible.

Note: If the core cutter can be pressed in gently, this is ideal, but if the core cutter will not penetrate the soil using this method, strike the core cutter head with a hammer to drive it into the soil. During this process, rather than using a light hammer and striking it many times with great force, it would be more effective to use a heavy hammer and strike it a few times in order to reduce the play between the core cutter and the soil. In addition, when a large diameter core cutter is used, it may penetrate the soil more easily if the soil around the outer wall of the core cutter is removed at the time of penetration.

c) Check to make sure soil fills the core cutter section and the core cutter has penetrated the soil sufficiently down to the core cutter head. Then use a scoop to dig up the core cutter and soil together.

Note: Caution is needed, since if the penetration depth exceeds the sum of the core cutter height and the height of the margin within the core cutter head, the soil inside the core cutter could be excessively compressed. When removing the core cutter and soil, if there is a hollow inside the core cutter at the plane of separation of the soil at the cutting edge, use a sample extruder or the like to press down the clump of soil inside the core cutter from top to bottom, with the excess soil on top of the core cutter remaining, to adjust so the hollow reaches the blade surface.

d) Remove the excess soil along both ends of the core cutter using the straight knife. Also remove the soil sticking to the outer wall of the core cutter and then use the scale to measure the mass m_2 (g).

Note: When shaping the soil inside the core cutter that has been excavated, observe the soil to make sure it does not include rock fragments, wood fragments or other foreign matter. In addition, if the blade has been significantly damaged as a result of being driven in, the collected soil may have become disturbed so it differs from the status at the site.

e) Press the sample out from the core cutter using an extruder, or remove it using a spatula or the like, and determine the water content w (%) using the method specified in JIS A 1203.

6 Test results

6.1 Wet density of soil

The wet density of the soil is calculated using equation (2) as follows.

$$\rho_t = \frac{m_2 - m_1}{V} \qquad (2)$$

where

ρ_t: Wet density of soil (g/cm³)
m_1: Mass of core cutter (g)
m_2: Mass of core cutter and soil (g)
V: Volume of core cutter (cm³)

6.2 Dry density of soil

Calculate the dry density of the soil using equation (3) below.

$$\rho_d = \frac{\rho_t}{1+(w/100)} \quad (3)$$

where

ρ_d: Dry density of soil (g/cm³)
w: Water content (%)

7 Reporting

The following items shall be reported as test results.

a) Site number and location

b) Test date

c) Test personnel

d) Core cutter specifications

e) Water content (%)

f) Wet density (g/cm³)

g) Dry density (g/cm³)

h) If the method used deviates in any way from this standard, give details of the method used.

i) Other reportable matters

Note: If the collected soil contains wood fragments or other foreign matter, this fact shall be reported.

Note: If another tool is used in place of the core cutter, the type of tool shall be reported.

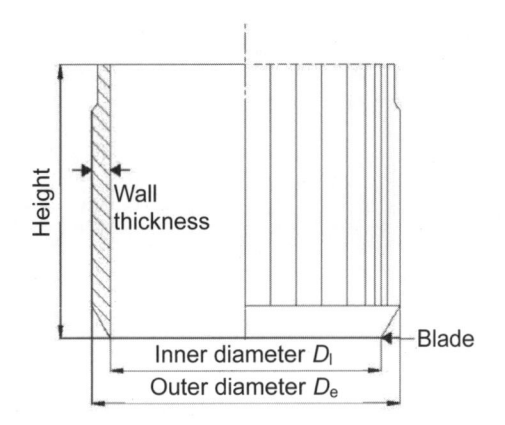

Fig. 1 Example of core cutter

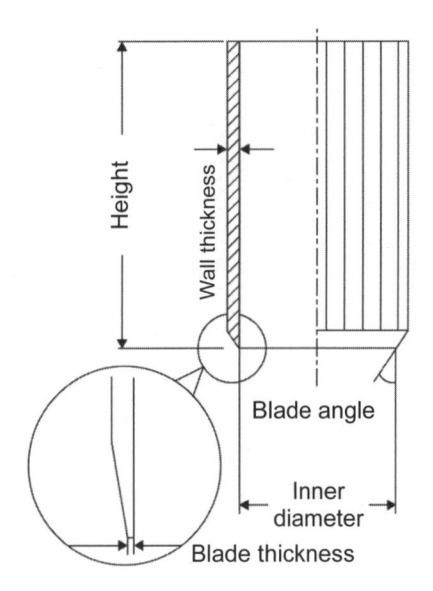

Fig. 2 Example of use of thin wall sampling tube as core cutter

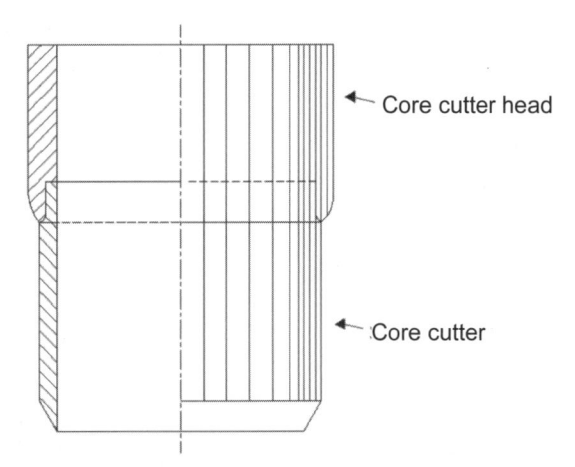

Fig. 3 Example showing coupled core cutter and core cutter head

Japanese Geotechnical Society Standard (JGS 1614-2012)
Test method for soil density using nuclear gauge

1 Scope

This standard specifies methods for conducting tests to determine the density and water content of the soil in natural ground and improved ground, through the use of a gauge that uses radioisotopes. This standard applies to all soil materials.

2 Normative references

The following standard shall constitute a part of this standard by virtue of being referenced herein. The latest version of this standard shall apply (including supplements).

JIS A 1203   Test method for water content of soils

3 Terms and definitions

The main terms and definitions used in this standard are as follows:

3.1 Radioisotope gauge

A radioisotope gauge is an instrument equipped with gamma-ray densitometers and neutron moisture gauges that use radioisotopes (RI) to measure wet density and water content.

3.2 Standard body

A standard body is a uniform substance composed of acrylic, lead and other materials that are physically and chemically stable and do not change over time.

3.3 Background radiation

Background radiation is natural radiation from cosmic rays and emitted from the ground.

3.4 Count rate

The count rate is the number of gamma ray and neutron radiation counts per minute.

Note: The count rate is expressed as counts per minute (cpm).

3.5 Standard body count rate

The standard body count rate is the count rate measured when the radioisotope gauge is placed on top of the standard body.

3.6 Standard body background radiation count rate

The standard body background radiation count rate is the count rate when the radioisotope gauge is placed on top of the standard body and the background radiation is measured.

3.7 Count rate ratio

The count rate ratio is the value determined by subtracting the standard body count rate from the count rates for the wet density and water content measured for the soil.

4 Equipment

4.1 Radioisotope gauge

There are two types of radioisotope gauge, a direct transmission type and a backscatter type. The type is chosen in accordance with the measurement method that is used. Fig. 1 and Fig. 2 show the abbreviated configuration of these radioisotope gauges. The radioisotope gauge shall be capable of measuring the wet density and water content simultaneously, and it shall be capable of measuring the background radiation and correcting the wet density and water content values. The radioisotope gauge comprises the following instruments.

a) Radiation source

The radiation source is a sealed unit containing a radioisotope that is strong enough to allow effective measurement of density and water content. Gamma rays are used as the radiation source for measuring density, while neutron radiation is used as the radiation source for measuring water content. The "Act Concerning Prevention from Radiation Hazards due to Radioisotopes, etc." (Radiation Hazard Prevention Act) and related laws and regulations shall be observed in the handling of radiation sources.

In addition, radioisotope gauges that contain radiation sources shall be those that have been approved under the Radiation Hazard Prevention Act and bear the seal of approval attesting to that fact.

b) Radiation source rod

The radiation source rod shall be a metal rod to which the radiation source has been attached. It shall have a configuration that enables it to be inserted into a hole that has been excavated in advance, and shall be a sturdy one that can be maintained securely at the specified depth.

The radiation source rod for a direct transmission type radioisotope gauge shall be a detachable one that is mounted inside the radioisotope gauge.

c) Detector

The detector shall be able to detect radiation in a stable manner.

d) Counter

The counter shall be an apparatus that can count the signals from the detector. The counter shall comprise a counting circuit, a timer circuit and other units.

e) Display

The display shall be able to display the measurements of radiation count rate, density, water content and so on.

f) Power supply

The power supply shall be equipped with the batteries needed to operate the instruments and shall be rechargeable.

Note: In the case of a scattering type radioisotope gauge, measurements taken in soil with a high gravel content may fluctuate considerably.

4.2 Standard body

The standard body shall be a substance composed of acrylic, lead or other material that is physically and chemically stable and does not change over time.

Note: The objective of using a standard body is to correct the attenuation over time in the strength of the radiation emitted by the radiation source. It is also used to confirm that the instrument is operating normally.

4.3 Ground shaping apparatus

The ground shaping apparatus shall be a steel plate, straight knife or other apparatus that is needed to smooth the surface of the ground at the measurement location so it is level.

4.4 Measurement hole preparation apparatus

The measurement hole preparation apparatus is needed to prepare a hole vertically into the ground that is to be measured when a direct transmission type radioisotope gauge is used. It shall comprise a guide plate and a hammer and drive rod or drill. The diameter of the drive rod or drill shall be the same as that of the radiation source rod.

4.5 Calibration container

The calibration container shall be sufficient to encompass the effective volume of the radioisotope gauge, and shall have a shape that enables the soil to be evenly compacted inside the container.

4.6 Other equipment

Other equipment shall be as follows.

a) Water content measurement apparatus

   The water content measurement apparatus shall be as specified in JIS A 1203.

b) Compacting apparatus

   The compacting apparatus shall be an apparatus that can compact the soil statically.

5 Test method

5.1 Establishment of reference values

Incorporate the radiation source into the radioisotope gauge and set the standard body count rate S_1 (cpm) of the day (reference date) on which the radiation source intensity was measured as the reference value of the radioisotope gauge.

The reference value shall be established as follows.

a) If another radiation source is present, conduct measurements far enough away from that radiation source to ensure that the measurements will not be affected.

b) Turn on the power to the instrument and allow it to warm up in order to stabilize the instrument.

c) Remove the radiation source rod from the radioisotope gauge and, at a distance of 20 m or more from the radiation source rod, measure the standard body background radiation on the standard body.

d) Place the standard body on the surface of concrete, asphalt or soil and, with the radiation source rod mounted in the radioisotope gauge, place the radioisotope gauge on the standard body and measure the standard body count rate. The standard body shall be placed at least 1.5 m away from walls, objects or the like to ensure that it will not be affected by walls or the like.

   Conduct measurements until the conditions in Equation (1) below are satisfied. Record the standard body count rate S_1 (cpm) and measurement time t_1 (min) for the reference date.

   $$S_1 \cdot t_1 \geq 154000 \tag{1}$$

   where

   S_1: Standard body count rate on reference date (measured standard body count rate minus standard body background radiation count rate)

t_1: Measurement time (min.) required to achieve $S_1 \cdot t_1 \geq 154000$ on reference date

Note: The date on which the reference value was established shall be used as the reference date for determining the attenuation of radiation.

5.2 Check of initial operational status

Determine the attenuation in the intensity of radiation during the time that has elapsed from the reference date until the date on which the initial operation of the radioisotope gauge was checked (inspection date), and determine the estimated standard body count rate S_3 (cpm) on the inspection date. Compare this calculation with the standard body count rate obtained through the standard body measurement conducted on the inspection date, in order to check the initial operational status of the instrument.

The initial operational status shall be checked as follows.

a) Conduct the process in 5.1 a) - c).

b) Place the standard body on the surface of concrete, asphalt or soil and, with the radiation source rod mounted in the radioisotope gauge, place the radioisotope gauge on the standard body and measure the standard body count rate. Conduct measurements until the conditions in Equation (2) below are satisfied. Record the standard body count rate S_2 (cpm) and the measurement time t_2 (min) for the inspection date.

$$S_2 \cdot t_2 \geq 154000 \qquad (2)$$

where

S_2: Standard body count rate on inspection date (measured standard body count rate minus standard body background radiation count rate)

t_2: Measurement time (min.) required to achieve $S_2 \cdot t_2 \geq 154000$ on inspection date

Note: The degree in the attenuation of radiation intensity of the radiation source over time will be determined by the type of radioisotope used as the radiation source. Accordingly, it is possible to determine approximately what the standard body count rate will be on the date on which the initial operational check is performed, by using the number of days from the reference date to the date on which the initial operational check was performed.

5.3 Creation of calibration curve

Before using the radioisotope gauge at the site, a calibration curve (calibration equation) showing the relationship between the wet density and water content of the soil being measured and the count rate ratio shall be prepared as follows.

a) Prepare the necessary number of specimens that have been evenly compacted to the specified density and water content inside the calibration container.

Note: Prepare the number of specimens needed to sufficiently encompass the range of ground density and water content values for the soil being measured, so a smooth curve can be obtained.

b) Determine the wet density ρ_t (g/cm³) for the individual specimens that have been prepared.

c) Conduct the procedure in 5.1 a) - c).

d) Place the standard body on the surface of concrete, asphalt or soil and, with the radiation source rod mounted in the radioisotope gauge, place the radioisotope gauge on the standard body and measure the standard body count rate. Determine the attenuation in radiation from the reference date to the date of calibration and confirm the status of operation using the same method as in 5.2 b).

e) Measure the background radiation at the surface of the specimens in the calibration container. To measure the background radiation, remove the radiation source rod from the radioisotope gauge and conduct measurements at a distance of 20 m or more.

f) Use a straight knife or the like to remove any protrusions from the top surface of the specimen so an area of the necessary size is smooth, in order to ensure that there is sufficiently close contact between the radioisotope gauge and the top surface of the specimen. In the case of a direct transmission type radioisotope gauge, use a measurement hole excavation apparatus to prepare a hole vertically into the specimen into which the radiation source rod can be inserted.

Note: In the case of a scattering type radioisotope gauge, undulations in the surface of the ground being measured will have a greater impact on measurements than in the case of a transmission type radioisotope gauge, so make sure the ground surface is sufficiently smooth.

g) Use the radioisotope gauge to measure the individual specimens and determine the individual values for count rate ratio.

h) Measure the water content ratio w (%) of the individual specimens and determine the individual values for water content ρ_m (g/cm^3). For the water content of the specimens, after measuring the count rate ratio of the specimens, take samples from several locations when the specimens are demolished and conduct measurements using the method specified in JIS A 1203.

i) Determine the relationship between the count rate ratio and the wet density or water content. This relational expression is called a calibration curve.

Note: The calibration curve (calibration equation) is determined from the count rate ratio that has been corrected for background radiation.

5.4 Site measurements

Density and water content at the site shall be measured as follows.

a) Conduct the procedure in 5.1 a) - c) and 5.3 d).

b) Measure the background radiation at the measurement site. To do this, remove the radiation source rod from the radioisotope gauge and, at a location at least 20 m away, place the instrument on the surface of the soil and conduct the measurement. The site background radiation shall be measured each time the soil or the measurement date changes.

c) Use a straight knife or the like to remove any protrusions from the ground at the measurement location so an area of the necessary size is smooth, in order to ensure that there is sufficiently close contact between the radioisotope gauge and the ground surface. In the case of a transmission type radioisotope gauge, use a measurement hole preparation apparatus to prepare a hole vertically into the ground into which the radiation source rod can be inserted.

Note: In the case of a backscatter type radioisotope gauge, undulations in the surface of the ground being measured will have a greater impact on measurements than in the case of a direct transmission type radioisotope gauge, so make sure the ground surface is sufficiently smooth.

d) Attach the radiation source rod to the radioisotope gauge and place the radioisotope gauge at the measurement location. Measure and record the individual radiation count rate values.

Note: If an instrument that can print the results of site measurement is being used, attach the results to the specified form used to record the date and location of site measurement and other measurement conditions, and store these records for future use.

e) Subsequently, move to the next measurement point within the same area and repeat c) and d) the necessary number of times.

6 Test results

6.1 Standard body count rate during initial operational check

The standard body count rate during the initial operational check shall be as follows.

a) From the results of measurement of the standard body count rate on the inspection date, calculate the maximum standard body count rate and the minimum standard body count rate, using equation (3) below.

$$\left.\begin{array}{c}S_{2max}\\S_{2min}\end{array}\right\} = S_2 \pm 1.96\sqrt{\frac{S_2}{t_2}} \qquad (3)$$

where

S_{2max} : Maximum standard body count rate on inspection date (cpm)
S_{2min} : Minimum standard body count rate on inspection date (cpm)

b) From the results of measurement of the standard body count rate on the reference date, calculate the maximum estimated standard body count rate and minimum estimated standard body count rate on the inspection date, using equation (4) below.

$$\left.\begin{array}{c}S_{3max}\\S_{3min}\end{array}\right\} = S_3 \pm 1.96\sqrt{\frac{S_3}{t_1}} \qquad (4)$$

$$S_3 = S_1 \left(\frac{1}{2}\right)^{D/T}$$

where

$S_{3\,max}$: Maximum estimated standard body count rate on inspection date (cpm)
S_{3min} : Minimum estimated standard body count rate on inspection date (cpm)
S_3 : Estimated standard body count rate on inspection date (cpm)
T : Half-life of radioisotope mounted (d)
D : Elapsed time from reference date to inspection date (d)

Note: The half-life of the radioisotope shall be as follows.

Gamma rays: Cobalt 60 (^{60}Co): 1924 days
Neutron radiation: Californium252 (^{252}Cf): 964 days

c) If the values for S_{2max} and S_{2min} obtained in a) satisfy the following, the radioisotope gauge is operating normally.

$$S_{2max} > S_{3min}$$

$$S_{2min} < S_{3max}$$

6.2 Density at time of calibration

The density at the time of calibration shall be as follows.

a) Calculate the wet density of the individual specimens using equation (5) below.

$$\rho_t = \frac{m}{V} \qquad (5)$$

where

ρ_t: Wet density (g/cm³)
m: Mass of specimen (g)
V: Volume of specimen (cm³)

b) Calculate the water content of the individual specimens using equation (6) below.

$$\rho_m = \frac{w/100}{1+w/100} \times \rho_t \tag{6}$$

where

ρ_m: Water content (g/cm³)
w: Water content ratio of specimen (%)

6.3 Density and water content of soil at site

Calculate the density and water content of the soil at the site as follows.

a) From the individual values for radiation count rate ratio obtained through measurement, read the wet density and water content according to the calibration curve.

b) Calculate the dry density for the measurement location using equation (7) below.

$$\rho_d = \rho_t - \rho_m \tag{7}$$

where

ρ_d: Dry density (g/cm³)

c) Calculate the water content ratio at the measurement location using equation (8) below.

$$w = \frac{\rho_m}{\rho_t - \rho_m} \times 100 = \frac{\rho_m}{\rho_d} \times 100 \tag{8}$$

where

w: Water content ratio (%)

7 Reporting

The following items shall be reported.

a) Type and instrument number of radioisotope gauge used
b) Water content ratio (%)
c) Wet density (g/cm³)
d) Dry density (g/cm³)
e) If the method used deviates in any way from this standard, give details of the method used.
f) Other reportable matters

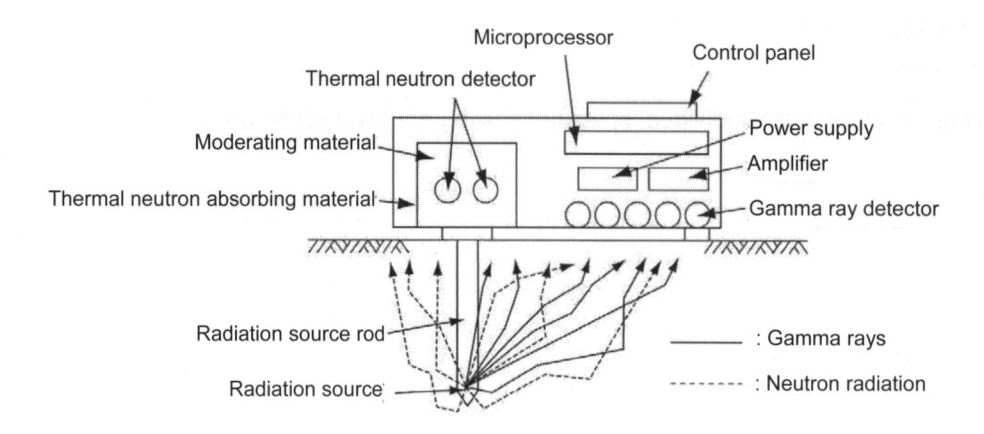

Fig. 1 Abbreviated configuration of direct transmission type radioisotope gauge

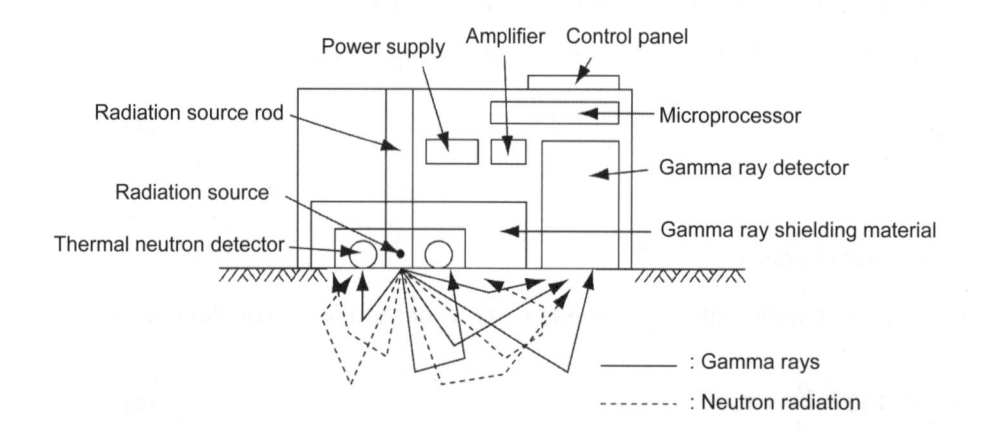

Fig. 2 Abbreviated configuration of backscatter type radioisotope gauge

Japanese Geotechnical Society Standard (JGS 1711-2012)
Method for measuring displacement of ground surface using stakes

1 Scope

This standard specifies methods for obtaining the vertical and horizontal displacement of the ground surface due to embankment or excavation using stakes. This standard applies primarily to natural ground and improved ground.

2 Normative references

The following standards shall constitute a part of this standard by virtue of being referenced in this standard. The latest versions of these standards shall apply (including supplements).

JIS B 7912-2   Field procedures for testing geodetic and surveying instruments -- Part 2: Levels
JIS B 7912-4   Field procedures for testing geodetic and surveying instruments -- Part 4: Electro-optical distance meters

3 Terms and definitions

The main terms and definitions used in this standard are as follows:

3.1 Vertical displacement

Vertical displacement is the amount of displacement obtained from the change in the difference in height between displacement stakes and reference stakes placed on the ground surface.

3.2 Horizontal displacement

Horizontal displacement is the amount of displacement obtained from the change in horizontal distance between displacement stakes and reference stakes placed on the ground surface.

4 Measurement apparatus

The measurement apparatus shall be as follows.

4.1 Displacement stake

A displacement stake is a stake with a nail driven into its head, which is used as a marker.

Example: Fig. 1 shows an example of a displacement stake. Normally a wooden stake measuring 100 - 150 mm square or with a top end measuring 100 - 150 mm and a length of 1.0 - 1.5 m is used as the displacement stake.

4.2 Reference stake

A reference stake is a stake that does not become displaced during the period of measurement and can serve as a reference point.

Note 1: Structures may be used as reference points for measuring vertical and horizontal displacement.

Note 2: If the reference stakes are to be used for a long period of time, stone pillars or concrete piles with markers on the top should be used.

Note 3: If the reference stakes are to be used only for a short period of time, wooden stakes with iron spikes or nails driven into their heads may be used.

4.3 Measuring instrument

The measuring instrument shall be as follows.

a) A level and a staff shall be used for measuring vertical displacement.

　　The level shall be one whose performance satisfies the requirements of JIS B 7912-2 (Field procedures for testing geodetic and surveying instruments -- Part 2: Levels).

b) Steel tape or an electro-optical distance meter and prism shall be used for measuring horizontal displacement. An appropriate measuring instrument shall be selected with consideration for the measurement location, measurement period, measurement range, required accuracy and so on.

　　The electro-optical distance meter shall be one whose performance satisfies the requirements of JIS B 7912-4 (Field procedures for testing geodetic and surveying instruments -- Part 4: Electro-optical distance meters).

5 Measurement method

5.1 Placement of displacement stakes and reference stakes

The displacement stakes and reference stakes shall be placed as follows.

Example: Fig. 2 shows an example of the placement of the displacement stakes and reference stakes.

a) The displacement stakes shall be placed so they are as vertical as possible. The tops of the stakes shall be placed so they are level and in a straight line, in such a way that they are unaffected by irregularities in the ground surface, and they shall be driven into the ground securely. If the surface of the ground is soft, they shall be driven as deep into the ground as possible.

Note 1: The height of the stake from the ground surface should be 200 - 500 mm.

Note 2: The placement of displacement stakes should be determined based on the ground conditions, embankment conditions, site conditions and so on. Normally they shall be placed at 2 - 3 m intervals in a straight line from the toe of the embankment slope.

Note 3: A monitoring platform should be set up if there is a possibility that the displacement stakes may be moved due to the measurement work.

b) The reference stakes shall be placed at a location where they will not be affected by work vehicles or passers-by, and where the reference stakes themselves will not be affected by settling or horizontal displacement.

Note 4: If the reference stakes will be used for a long period of time, their foundations should be cast in concrete.

5.2 Measurement of vertical and horizontal displacement

Vertical and horizontal displacement shall be measured as follows.

a) Vertical displacement shall be measured as follows.

　　1) Vertical displacement shall be obtained from the change in the difference in height between the displacement stakes and the reference stakes.

　　2) The change in the difference in height between the displacement stakes and the reference stakes shall be measured using a level, by placing a staff on the top of the displacement stake and measuring in 1 mm increments.

b) Horizontal displacement shall be measured as follows.

   1) Horizontal displacement shall be obtained from the change in horizontal distance between the displacement stakes and the reference stakes.

   2) The horizontal distance between the displacement stakes and the reference stakes shall be measured in 1 mm increments using steel tape or an electro-optical distance meter. If an electro-optical distance meter is used, a prism shall be placed on the top of the displacement stakes and measurement shall be performed.

c) Immediately after the displacement stakes and reference stakes have been placed, measure the mutual difference in height and the horizontal distance and use those values as initial values.

d) Repeat the measurements in a) and b) after each passage of the specified number of days, and calculate the difference between the measurements and the initial values to obtain the amount of displacement.

6 Test results

The amount of displacement is calculated by the following equation. The amount of vertical displacement is defined as positive in the direction of uplift. In the case of an embankment, the horizontal displacement is positive in the direction away from the embankment, and in the case of an excavation, the direction toward the excavation side is positive.

$$\Delta H = H - H_0$$

where

ΔH : Displacement (mm)
H : Measurement value at arbitrary measurement time (mm)
H_0 : Initial value (mm)

Note: The embankment fill height, excavation depth and other changes over time should also be noted in drawings.

7 Reporting

The following items shall be reported.

a) Date of displacement stake placement and placement location

b) Specifications of displacement stakes

c) Location of reference stake placement and placement method

d) Relationship between vertical and horizontal displacement and elapsed time

e) If the method used deviates in any way from this standard, give details of the method used.

f) Other reportable matters

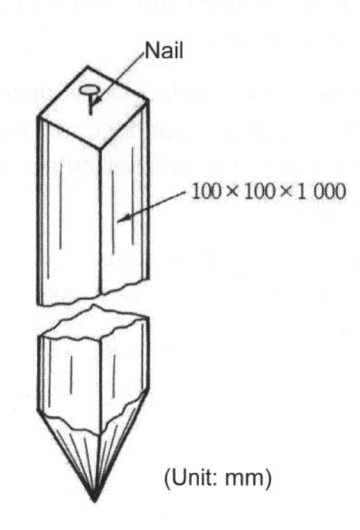

Fig. 1 Example of displacement stake

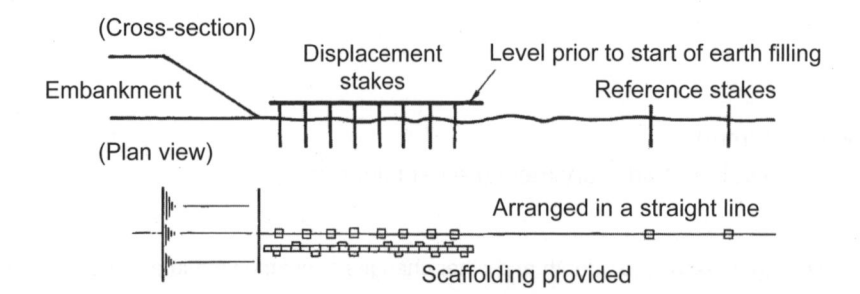

Fig. 2 Example of placement of displacement stakes and reference stakes in the case of an embankment

Japanese Geotechnical Society Standard (JGS 1712-2012)
Method for measuring settlement of ground surface using settlement plate

1 Scope

This standard specifies methods for measuring the settlement of the ground surface due to earth fill, landfill and so on, using a settlement plate.

2 Normative references

The following standard shall constitute a part of this standard by virtue of being referenced herein.

The latest version of this standard shall apply (including supplements).

JIS B 7912-2   Field procedures for testing geodetic and surveying instruments -- Part 2: Levels

3 Terms and definitions

The main terms and definitions used in this standard are as follows:

3.1 Ground surface settlement

The amount of settlement as determined by the change in the difference in levels between a settlement plate placed on the ground surface and reference pile that is regarded as stationary.

4 Measurement apparatus

The measurement apparatus shall be as follows:

4.1 Settlement plate

The settlement plate shall comprise a steel plate placed on the ground surface and a measuring rod connected perpendicularly to the steel plate. A protective pipe shall be provided around the outside of the measuring rod, forming a dual structure, so that the rod will not be affected by friction with the surrounding ground.

Note: Fig. 1 shows an example of a settlement plate. It is desirable that the size of the settlement plate is increased when the maximum grain size of the earth fill material is large, and when the fill height is high.

4.2 Steel plate

Generally, the steel plate shall be a square plate of 9 mm or more in thickness with each side 400 mm or more.

4.3 Measuring rod

The measuring rod shall be a steel rod of 12 mm diameter or more, with a measurement point on the top end. The bottom end of the measuring rod shall be welded to the steel plate. The top of the rod shall be externally threaded and connected to the rod in accordance with the earth fill height. The rod length shall be (1000±1) mm, and it shall be possible to correctly ensure the actual length of the rod.

4.4 Reference piles

The reference piles shall be the place at which the ground height is known and that can be regarded as reference points when settlement is measured. They shall be placed at locations where no displacement will occur over a long period of time.

Note 1: If the reference piles are to be used for a long period of time, stone pillars or concrete piles with measurement points on the top should be used, and their foundations should be fixed using concrete.

Note 2: Reference piles and reference points may be placed on the top of structures.

Note 3: If the reference piles are to be used only for a short period of time, wooden stakes with iron spikes or nails driven into their heads may be used.

4.5 Measuring instruments

A tracking level and a staff shall be used as the measuring instrument.

The level shall be one whose performance satisfies the requirements of JIS B 7912-2 (Field procedures for testing geodetic and surveying instruments -- Part 2: Levels).

5 Measurement method

5.1 Preparations

The preparations for measurement shall be as follows:

a) To improve the contact of the steel plate and the ground surface, the placement surface shall be thoroughly made smooth so that it is level, and the settlement plate shall be placed so that the steel plate is horizontal.

Note 1: To improve the contact between the steel plate and the ground surface, sand should be spread on top of the placement surface.

Note 2: When placing the steel plate in a sand mat, the steel plate should be placed at least 50 - 100 mm above the surface of the original ground, as shown in Fig. 2.

b) A measuring rod shall be connected vertically to the measuring rod that was previously set and the rods shall be protected so that they will not be bent during the earth fill process.

Note 3: The measuring rod that is assembled vertically during the earth fill process should be protected by placing a fence around the rod and so on, and the position of the measuring rod should be clearly indicated.

Note 4: The protective pipe should be installed with sufficient margin during settlement (500 mm or more with respect to an earth fill height of 10 m) between the pipe and the steel plate, in order to ensure that the protective pipe does not press the settlement plate downward as a result of compression by the earth fill.

c) A small compacting machine (compactor, etc.) shall be used to compact the area around the measuring rod so its stiffness matches the rigidity of the earth fill as closely as possible.

d) The reference piles shall be placed at a location where they will not be affected by vehicles or passers-by, and where the reference piles themselves will not be affected by settlement or other ground movements.

5.2 Measurement

Measurement shall be conducted as follows:

a) Place a staff vertically on the top of the reference pile and use the tracking level to read the value on the staff in 1 mm precision.

b) Place the staff vertically on the top end of the measuring rod and use the tracking level to read the value on the staff in 1 mm precision.

c) The difference between the reading on the measuring rod when the settlement plate has stabilized after placement, and the reading on the reference pile, shall be used as the initial difference in height.

d) Conduct the measurements in a) and b) for each specified period of time, and determine the difference in height between the reference pile and the measuring rod.

e) Determine the amount of ground surface settlement from the initial difference in height determined in c) and the height difference determined in d).

Note: The frequency of the measurements should be determined based on the ground conditions, the importance and progress of the construction work, and the speed of ground settlement and so on.

f) When a new measuring rod is connected to the existing measuring rod, the extended length of the measuring rod shall be added to the initial difference in height in order to update the value for the initial difference in height.

6 Test results

The ground surface settlement after the start of measurement shall be calculated using the following equation.

$$\Delta h = h - h_0$$

where

Δh : Ground surface settlement (mm)

h : Difference in height between reference pile and measuring rod at the time of an each measurement (mm)

h_0 : Initial difference in height (initial difference in height between immobile stake and measuring rod) (mm)

7 Reporting

The following items shall be reported:

a) Date of the settlement plate installation and the location of placement

b) Specification of the settlement plate

c) Location of the reference pile and the method of placement

d) Relationship between the amount of settlement and the elapsed time

e) If a method that partially differs from this standard was used, details of the points of difference

f) Other reportable matters

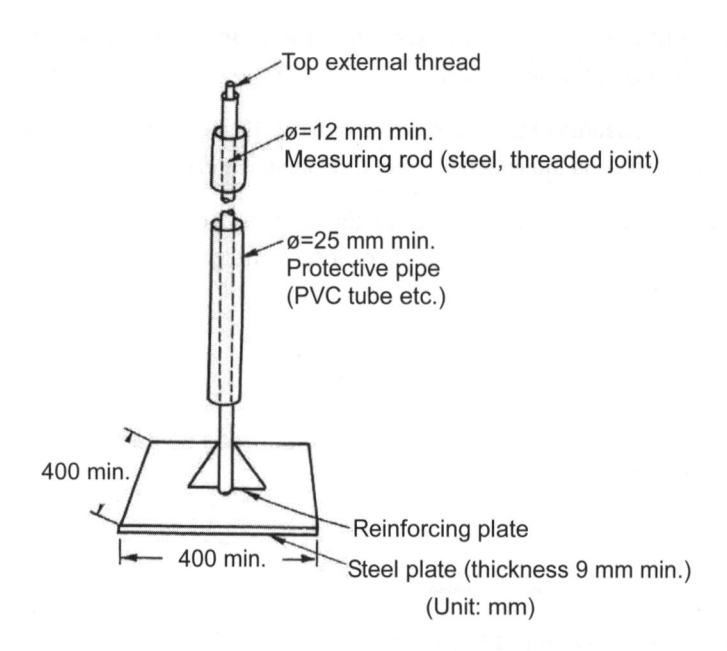

Fig. 1 Example of settlement plate

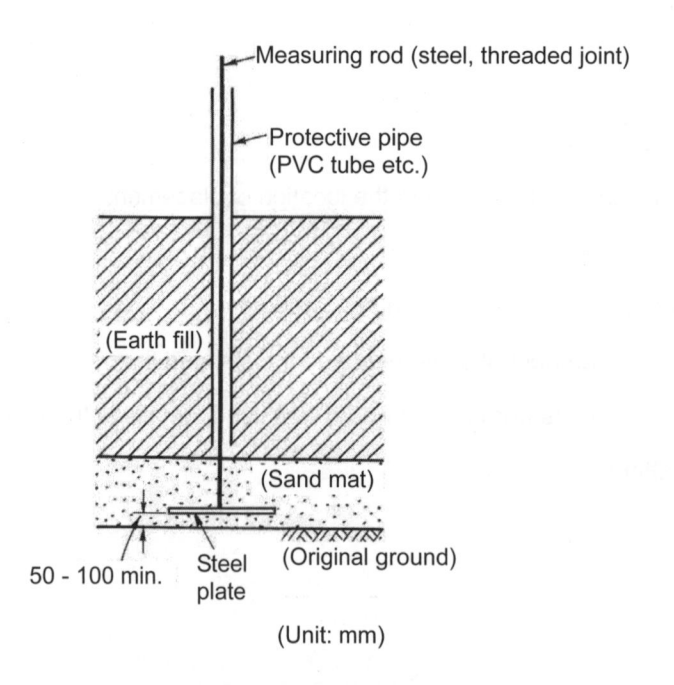

Fig. 2 Example of placement of settlement plate in sand mat

Japanese Geotechnical Society Standard (JGS 1718-2012)
Method for measuring vertical displacement of embankment using cross arm settlement gauge

1 Scope

This standard specifies methods for determining the amount of vertical displacement in an embankment in-situ, using a crossarm settlement gauge. This method applies primarily to high embankments.

Note: In this standard, high embankment refers to an embankment with a height of approximately 15 m or more.

2 Normative references

The following standard shall constitute a part of this standard by virtue of being referenced herein. The latest version of this standard shall apply (including supplements).

JIS B 7912-2   Field procedures for testing geodetic and surveying instruments -- Part 2: Levels

3 Terms and definitions

The main terms and definitions used in this standard are as follows:

3.1 Vertical displacement in embankment

The amount of displacement measured using a crossarm settlement gauge placed in the depth direction in an embankment.

4 Measurement equipment

The measurement equipment shall be as follows.

4.1 Crossarm settlement gauge

The crossarm settlement gauge shall comprise crossarms, vertical inner pipes, vertical outer pipes, a foundation anchoring steel pipe, and a detector section.

a) Crossarm

   The crossarm shall be a steel member with a grooved cross-section for placement in the embankment. Its measurements shall be approximately 2,000 mm in length by 80 mm in height by 40 mm in width. The crossarms shall have sufficient strength and the ability to track vertical displacement in the embankment.

b) Vertical inner pipes

   The vertical inner pipes shall be steel pipes that are rigidly connected vertically to the crossarms. Their measurements shall be approximately 50 mm in diameter and 1,000 - 1,800 mm in length.

Note: Hard PVC pipes may be used if they will pose no barriers to measurement, for example when the embankment height is low.

c) Vertical outer pipes

   The vertical outer pipes shall be steel pipes or hard PVC pipes that are placed in the interval between the vertical inner pipes above and below, in such a manner that they are able to slide. Their measurements shall be approximately 60 mm in diameter and 2,000 mm in length.

d) Foundation anchoring steel pipe

The foundation anchoring steel pipe shall be a steel pipe with a bottom cover that is fastened to a reference point that has been placed in the foundation ground. The diameter shall be approximately equal to that of the vertical outer pipes.

e) Detector section

The detector section shall be able to measure the vertical position of each crossarm in 1 mm increments. The following types shall be available.

1) In the mechanical detector, a torpedo is inserted manually from the hole opening as shown in Fig. 1. When the steel tube connected to the torpedo is pulled, the spring on the torpedo opens and engages the lower ends of the vertical inner pipes to detect the position of the crossarm.

2) In the electromagnetic detector (manual type), a search element is inserted manually from the hole opening as shown in Fig. 2, and the position of the crossarms is detected magnetically.

3) In the electromagnetic detector (automatic type), the amount of movement of the crossarms is detected using an electrical settlement gauge attached to a measurement rod whose lowest part is a fixed point, as shown in Fig. 3.

4.2 Fixed stakes

The fixed stakes shall be able to serve as reference points for vertical displacement measurement.

Note: If the fixed stakes are to be used for a long period of time, stone pillars or concrete piles with measurement points on the top should be used, and their foundations should be consolidated using concrete.

4.3 Measuring instruments

A level and staff shall be used as the measuring instruments.

The level shall be one whose performance satisfies the requirements of JIS B 7912-2 (Field procedures for testing geodetic and surveying instruments -- Part 2: Levels).

5 Measurement method

The measurement method shall be as follows.

5.1 Placement of crossarm settlement gauge and fixed stakes

a) Excavate a hole measuring approximately 100 mm in diameter and 2,000 mm in depth at the specified location and place the foundation anchoring steel pipe vertically into the hole. Then fill the surrounding area with an appropriate material.

Note 1: Cement slurry, concrete or other materials can be used as fill materials. The fill material should be determined based on the measurement conditions, geological conditions and so on.

b) The crossarm settlement gauge placement location and the number of gauges to be placed shall be determined as appropriate in accordance with the purpose of measurement. The vertical outer pipes and vertical inner pipes shall be placed so that they are able to slide freely, and their connecting sections shall be sealed to keep out soil. The crossarms shall be placed horizontally at the necessary positions in accordance with the progress of the embankment process. In addition, the vertical inner pipes and vertical outer pipes shall be placed so that they remain vertical and shall be protected so that they do not bend during the embankment construction.

Note 2: Rubber sleeves, linen tape, butyl rubber and so on can be used as sealing materials.

c) The area around the vertical outer pipes shall be compacted using a small compacting machine (compactor, etc.) so that its stiffness matches the stiffness of the surrounding embankment as closely as possible.

d) The fixed stakes shall be placed at locations where they will not be affected by vehicles or passers-by, and where the fixed stakes themselves will not be affected by settlement of the embankment.

5.2 Measurement of vertical displacement

a) Each time a new crossarm is placed in the embankment, measure the height of that crossarm and the difference in height between the fixed stake and the top of the pipe protruding from the surface of the ground using a level or like. At the same time, measure the height of each of the crossarms that have been placed up to that time, in order from the top down using a detector.

b) Determine the difference in height between the fixed stakes and the crossarms at the time of crossarm placement and use this value as the crossarm initial location. Record the date as the date of the start of the embankment.

c) During the embankment construction and after the embankment has been completed, record the elapsed time, and use the detector to measure the locations of the crossarms.

d) Determine the cumulative vertical displacement of the crossarms by subtracting the measurement determined in c) from the initial crossarm location.

e) Measure the location of the crossarms as frequently as needed during the embankment construction process and after the embankment has been completed.

6 Analysis of results

The measurement results shall be analyzed as follows.

a) The cumulative vertical displacement of each crossarm for each change in elapsed time shall be determined as the initial default value at the time of placement of each crossarm, minus the subsequent measurements.

b) Prepare a diagram showing the relationship between the vertical displacement in the embankment and the elapsed time.

  1) Draw a graph with the elapsed time as the horizontal axis and the cumulative vertical displacement of each crossarm and the embankment height as the vertical axis, in order to show the changes in cumulative vertical displacement over time.

  2) Draw a graph with the cumulative vertical displacement of each crossarm as the horizontal axis and the initial location of each crossarm at the time of placement as the vertical axis, in order to show the cumulative vertical displacement distribution.

7 Reporting

The following items shall be reported.

a) Date of placement of crossarm settlement gauge

b) Diagrams showing horizontal position of crossarm settlement gauge and position in depth direction

c) Specifications of measurement apparatus

d) Date of measurement

e) Measurements and drawings showing the relationship between vertical displacement in the embankment and elapsed time

f) If a method that partially differs from this standard was used, details of the points of difference

g) Other reportable matters

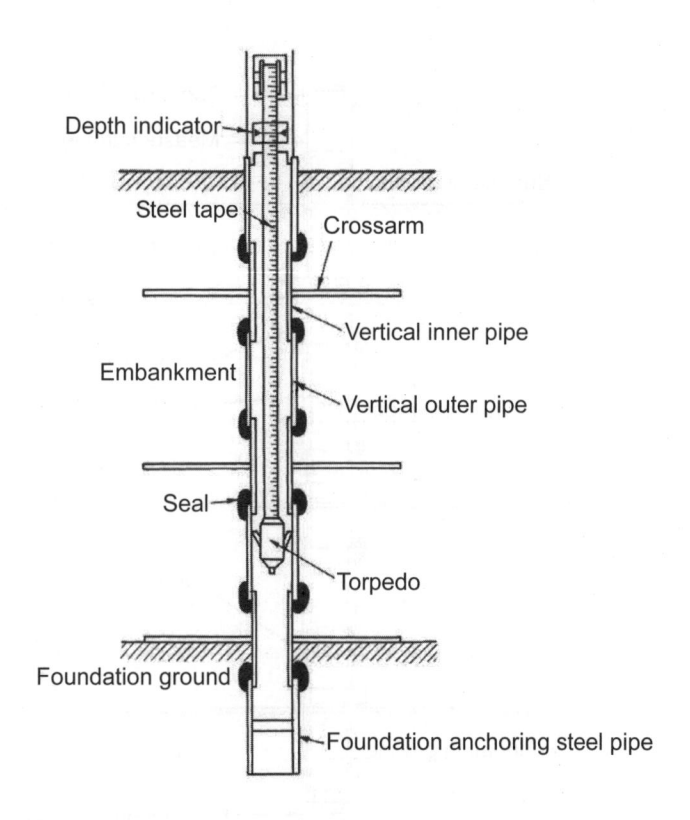

Fig. 1 Example of crossarm settlement gauge using mechanical detector

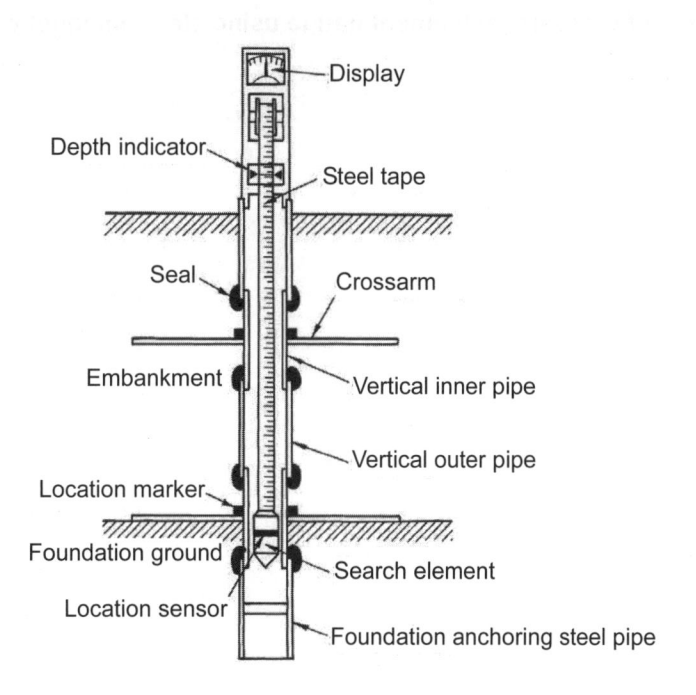

Fig. 2 Example of crossarm settlement gauge using electromagnetic detector (manual type)

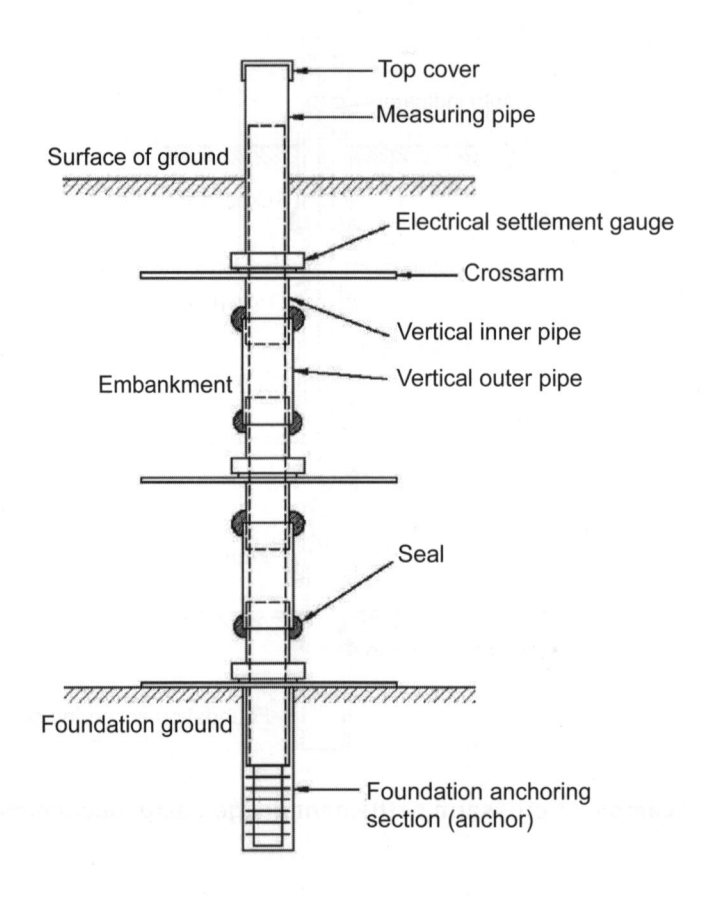

Fig. 3 Example of crossarm settlement gauge using electromagnetic detector (automatic type)

Japanese Geotechnical Society Standard (JGS 1721-2012)
Method for measuring tilt of ground surface using tiltmeter

1 Scope

This standard specifies methods for determining the variation over time in the inclination of the ground surface, using a water tube type tiltmeter. This standard applies primarily to natural ground, reclaimed ground and embankment.

2 Normative references

None

3 Terms and definitions

The main terms and definitions used in this standard are as follows:

3.1 Variation in inclination of ground surface

The change over time in the angle and direction of inclination.

4 Measurement apparatus

The measurement apparatus shall be as follows.

4.1 Water tube type tiltmeter for measuring ground inclination (hereafter "tiltmeter")

This apparatus shall have a structure in which a main bubble tube and secondary bubble tube are mounted on top of a bubble tube platform that is supported by a tripod (a main support and secondary supports on the left and right), and the inclination of the bubble tube platform can be adjusted by raising or lowering the tripod.

a) Main support

   A rotation plate is fastened to the main support so that, by rotating the main support, the angle of inclination in the main bubble tube direction can be converted based on the reading of the rotation plate.

b) Secondary supports

   The secondary supports are used to adjust the perpendicular direction of the main bubble tube and the horizontal position of the bubble tube platform.

c) Main bubble tube

   The main bubble tube shall be attached so it is at a right angle to the secondary bubble tube. The scale markings shall be at intervals of 2 mm or more and the angle of inclination of each scale marking shall be 12 seconds or less.

d) Rotation plate

   The rotation plate is marked with 360 scale markings around its circumference and has a measurement range of ±1° or more, so the angle of inclination for each scale marking is 2 seconds or less. The tiltmeter shall be calibrated prior to use to determine the calibration factor (the angle of inclination for each scale marking on the rotation plate). The liquid used inside the bubble tube shall not freeze even when the outside air temperature is -10 °C, and the change in viscosity due to variations in temperature shall be small.

Fig. 1 shows an example of a tiltmeter.

5 Measurement method

The measurement method shall be as follows.

5.1 Installation of measurement apparatus

The measurement apparatus shall be installed as follows.

a) The location selected for installation shall be one that is little affected by fluctuations that will interfere with the purpose of measurement.

Note 1: The apparatus should also be installed at a place that clearly seems to be a stationary point, and comparative observations should be conducted.

Note 2: The causes of interference include strong vibrations due to the passage of vehicles and so on, and tilting caused by the growth of tree roots.

b) After excavating the surface soil, place gravel and concrete to create a level surface, and then install the installation platform in an area measuring 500 x 500 mm or larger. The installation platform shall be a solid one that is sufficiently embedded in the ground or fastened in place to a foundation pile.

Fig. 2 shows an example of an installation platform.

Note 3: If wooden stakes are used as foundation piles, a wooden stake with a length of 1 m and a top end of 90 mm or larger shall be driven into the ground so the top of the stake is approximately 200 mm below the surface of the ground. If the foundation pile reaches bedrock before it has been driven in for the specified length and cannot be driven in further, the pile may be cut so the head of the pile is approximately 200 mm below the surface of the ground. If the ground surface is rock and foundation piles cannot be driven in, an installation platform may be fabricated without foundation piles.

c) The tiltmeters shall be installed as follows.

   1) Two tiltmeters shall be installed at right angles to one another on the installation platform. As a rule, the main bubble tube directions shall be N-S and E-W, and the main supports shall be on the N and E sides.

   2) As a rule, the tiltmeters shall be installed one week or more after casting the concrete for the installation platform.

   3) The tiltmeters shall be mounted on a glass plate covering the installation platform surface, or on pedestals attached at the positions of the main support and secondary supports.

d) A protective case shall be placed on top of the installation platform to protect the tiltmeter. The protective case shall have an internal configuration that can ventilate and drain water.

5.2 Measurement method

Measurements shall be conducted as follows.

a) The measurements to be conducted at the time of installation shall be as follows.

   1) Turn the secondary supports and adjust so the air bubble in the secondary bubble tube is in the center of the scale. Then turn the main support and adjust so the air bubble in the main bubble tube is in the center of the scale.

   2) Read the indicator value on the rotation plate after adjustment and use this as the initial value.

b) The measurement method shall be as follows

   1) Check the direction of inclination, record the position of the air bubble in the main bubble tube, and then turn the main support and adjust so the bubble is in the center.

   2) After adjustment, read the direction of rotation of the main support and the indicator value on the rotation plate. When measurements are conducted, check for misalignment of the leg shafts and cracking of the bubble tube, and inspect the position of the air bubble in the secondary bubble tube. Also check to make sure the previous rotation plate reading matches the current rotation plate reading, and investigate the cause if they are not in agreement.

c) The interval between measurement times shall be established in accordance with the status of ground fluctuations and the status of the area around the site.

6 Test results

Annex A shows a method for analyzing the results.

7 Reporting

The following items shall be reported.

a) Date of tiltmeter installation and installation location

b) Specifications of tiltmeter and installation platform

c) Variation in inclination (in N-S and E-W direction), maximum angle of inclination, cumulative variation in inclination and average daily variation in inclination

d) Drawing showing cumulative variation in inclination, drawing showing direction of cumulative variation in inclination, drawing showing maximum variation in angle of inclination

e) If the method used deviates in any way from this standard, give details of the method used.

f) Other reportable matters

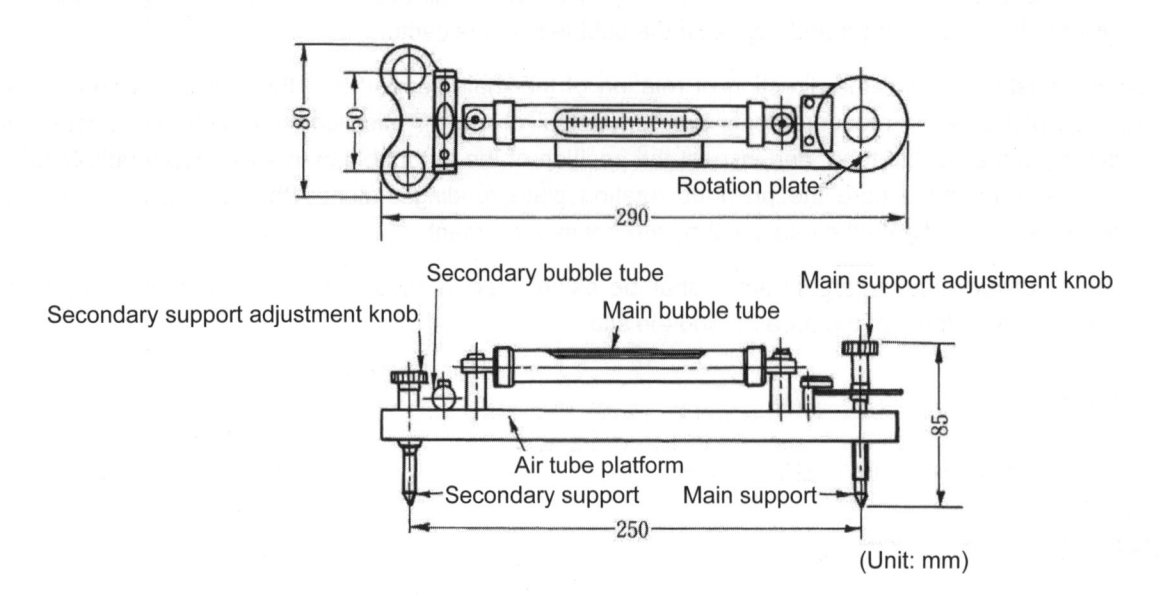

Fig. 1 Example of water tube type tiltmeter

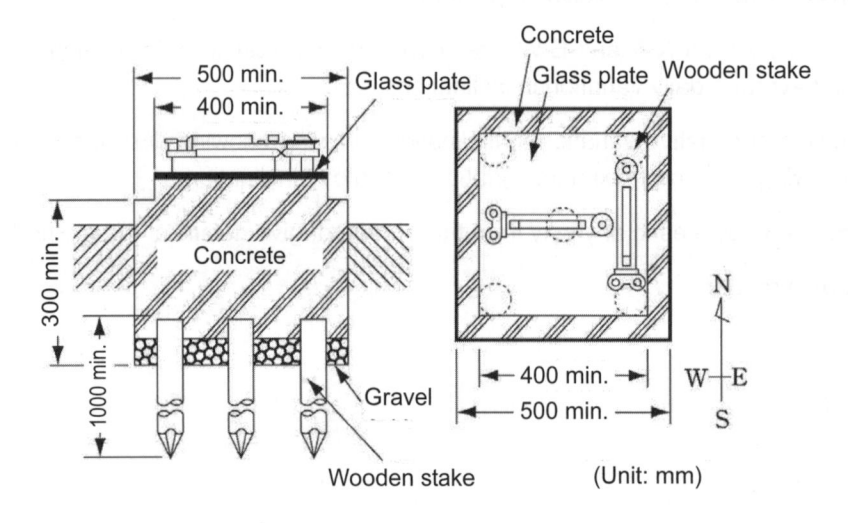

Fig. 2 Example of installation platform

Annex A
(Regulation)

Methods for analyzing results

A-1 Analysis of measurement results

The results of measurement shall be analyzed as follows.

a) Determine the elapsed number of days n from the date on which measurement was started.

b) Calculate the difference (x, y) from the previous rotation plate reading, using the following equation.

$$x = (\psi_2 - \psi_1) \times c$$

$$y = (\psi'_2 - \psi'_1) \times c$$

where

x: Fluctuation in N-S direction (seconds)
y: Fluctuation in E-W direction (seconds)
ψ_1: Previous reading in N-S direction (seconds)
ψ_2: Current reading in N-S direction (seconds)
ψ'_1: Previous reading in E-W direction (seconds)
ψ'_2: Current reading in E-W direction (seconds)
c: Calibration factor

c) Calculate the maximum angle of inclination, using the following equation.

$$\theta = \sqrt{x^2 + y^2}$$

where

θ: Maximum angle of inclination (seconds)
x: Difference from previous rotation plate reading in N-S direction (seconds)
y: Difference from previous rotation plate reading in E-W direction (seconds)

d) Calculate the average daily variation in angle of inclination, using the following equation.

$$\bar{\theta} = \frac{\Sigma \theta}{n}$$

where

$\bar{\theta}$: Average daily variation in angle of inclination (seconds)
θ: Maximum angle of inclination (seconds)
n: No. of days (days)

e) Calculate the inclined direction angle, using the following equation.

$$\Phi = \cos^{-1} \frac{|x|}{\sqrt{x^2 + y^2}}$$

where

Φ : Inclined direction angle (°)

x : Difference from previous rotation plate reading in N-S direction (seconds)

y : Difference from previous rotation plate reading in E-W direction (seconds)

f) Calculate the cumulative angle of inclination, using the following equation.

$$\Sigma R = \sqrt{X^2 + Y^2}$$

where

ΣR : Cumulative angle of inclination (seconds)

$X(=\Sigma_x)$: Cumulative variation in N-S direction (seconds)

$Y(=\Sigma_y)$: Cumulative variation in E-W direction (seconds)

g) Calculate the cumulative inclined direction angle, using the following equation.

$$\Sigma \Phi = \cos^{-1} \frac{|X|}{\sqrt{X^2 + Y^2}}$$

where

$\Sigma \Phi$: Cumulative inclined direction angle (°)

$X(=\Sigma_x)$: Cumulative variation in N-S direction (seconds)

$Y(=\Sigma_y)$: Cumulative variation in E-W direction (seconds)

Japanese Geotechnical Society Standard (JGS 1951-2012)
Method for air permeability test in vadose zone

1 Scope

This standard specifies methods for determining the permeability of gas in sandy, gravelly, or other soil in the vadose zone. The permeability of the gas in the soil determined using this method is the value for the moisture state during the test.

2 Normative references

None

3 Terms and definitions

The main terms and definitions used in this standard are as follows:

3.1 Vadose zone

The vadose zone is the ground above the level of groundwater that includes gas in the soil.

3.2 Permeability of gas in soil

A coefficient indicating the air permeability of the gas in the soil in the vadose zone.

3.3 Suction well

A well created for the purpose of sucking up the gas in the soil in the vadose zone.

3.4 Observation well

A well that is used to measure the pressure drop in the gas in the soil during the test.

3.5 Suction flow rate of gas in soil

The volume flow rate (at normal temperature and normal pressure) of the gas in the soil that is sucked up by the suction well.

3.6 Suction pressure

The pressure of the gas in the soil that acts on the suction well during the test.

3.7 Generated pressure

The pressure of the gas in the soil, determined by adding normal pressure to the pressure differential observed by the observation well during the test.

3.8 Normal temperature / normal pressure

The normal temperature shall be 20 °C and normal pressure shall be 101.325 kPa.

4 Equipment and apparatus

4.1 Drilling gear

The drilling gear comprises equipment that can drill a borehole of an arbitrary diameter down to an arbitrary depth without using drilling water.

4.2 Gas suction apparatus

The gas suction apparatus comprises equipment that can suck up the gas in the soil and whose suction pressure can be controlled.

Note: Suction pumps, blowers and other equipment can be used as gas suction apparatus. Pressure control for the gas suction apparatus can be achieved by providing a valve or an outside air intake port or the like.

4.3 Flowmeter

The flowmeter shall be able to measure the suction flow rate of the gas in the soil at an accuracy that is in keeping with the objectives of the test. The volume of the gas varies greatly depending on temperature and pressure, so the value shall be converted to the volume flow rate at normal temperature and normal pressure.

4.4 Pressure gauge

The pressure gauge shall be able to measure the absolute suction pressure, at an accuracy that is in keeping with the objectives of the test.

Note: A bourdon tube type and other types are available for use as the pressure gauge.

4.5 Differential pressure gauge

The differential pressure gauge shall be able to measure the difference in pressure as compared to absolute pressure, at an accuracy that is in keeping with the objectives of the test. The differential pressure gauge shall measure the drop in pressure at the observation well. Differential pressure is the difference in pressure between two points, and the differential pressure gauge shall be able to measure the difference in pressure between normal pressure and generated pressure.

4.6 Thermometer

The thermometer shall be an instrument that measures the temperature of the gas that is sucked up when the suction flow rate of the gas in the soil is measured.

4.7 Barometer

The barometer shall measure the atmospheric pressure at the test location. It shall be used for measurements to determine the pressure generated inside the observation well.

4.8 Pipe with screen

A pipe with holes or slits made in its surface. The pipe with screen shall be placed at the gas suction depth (in the case of the suction well) or at the pressure differential measurement depth (in the case of the observation well).

5 Air permeability test method

5.1 Placement of suction well

The suction well shall be placed by using the drilling gear to drill a borehole down to the specified depth and inserting the pipe with screen in such a manner that the screen of the suction well shall be at the specified depth. The gaps between the borehole and the pipe with screen shall be filled with filter material.

Consideration shall be given to the following when placing the suction well.

a) The groundwater level shall be measured in advance in order to determine the drilling depth for the suction well.

b) To prevent accidents due to hazardous gases, an investigation of the types and concentrations of the components of the gas in the soil shall be conducted in advance. If the gas that is sucked up is discharged, it shall be subjected to adsorption treatment using activated charcoal or the like.

c) The standard diameter of the suction well shall be inner diameter 25 - 50 mm. Airtight sealing material shall be filled in on top of the filter material to prevent air from the ground surface entering through the gaps between the borehole and the suction well pipe. The fill thickness of the airtight sealing material shall be approximately 1 m. Care shall be taken to ensure that the sealing material does not penetrate the filter layer during the sealing material fill process.

d) The screen hole area ratio shall be approximately 3 - 10%.

e) The filter material that is selected shall have an air permeability that is equal to or greater than that of the ground, and it shall allow as little as possible of the fine-grained particle content of the ground to enter the suction well.

f) The airtight sealing material that is selected shall not be deformed by the negative pressure produced by suction.

5.2 Placement of observation wells

The observation wells shall be placed by using the drilling gear to drill a borehole down to the specified depth and inserting the pipe with screen in such a manner that the screen of the observation well is at the depth of the center of the vadose zone. The gaps between the borehole and the pipe with screen shall be filled with filter material.

Consideration shall be given to the following when placing the observation wells.

a) The groundwater level shall be measured in advance in order to determine the location of the screen for the observation well.

b) The standard diameter of the observation well hole shall be inner diameter 25 - 50 mm. Airtight sealing material shall be filled in on top of the filter material to prevent air from the ground surface entering through the gaps between the borehole and the suction well pipe. The fill thickness of the airtight sealing material shall be approximately 1 m. Care shall be taken to ensure that the sealing material does not penetrate the filter layer during the sealing material fill process.

c) The filter material that is selected shall have an air permeability that is equal to or greater than that of the ground.

d) A pressure differential gauge, or a valve or the like that can be connected to a pressure differential gauge, shall be placed on top of the observation well, and shall be made airtight.

e) The screen shall have a length that enables the pressure generated at the specified depth to be measured.

f) Observation wells shall be placed at different distances from the suction well within the test zone. If possible, at least three observation wells shall be provided. Here "test zone" refers to the area of the vadose zone where the air permeability of the gas in the soil will be determined around the suction well.

Note: In many cases, the area within which sufficient observations are possible for a single suction well is approximately 10 m or less.

5.3 Test preparations

The test preparations shall be as follows.

a) Measure the distance from the center of the suction well to the centers of the observation wells.

b) Connect the gas suction apparatus, flowmeter and pressure gauge to the suction well.

c) Install the pressure differential gauges in the observation wells.

Note: Fig. A.1 in Annex A shows an example of the placement of equipment for air permeability tests at the site.

5.4 Constant flow rate suction test

The constant flow rate suction test shall be performed as follows.

a) Start suction at the specified gas suction flow rate. Regarding the suction flow rate, the suction pressure of the suction well shall be set so that a significant pressure change occurs in each observation well. When suction pressure is applied, the suction pressure shall be adjusted so a heterogeneous air flow (bypass flow) is not produced as a result of the application of excessive pressure, and so the test results do not reflect changes over time in the water content and air permeability of the ground due to the application of suction pressure.

Note: If the groundwater level is high, increase the distance between the suction well screen and the groundwater level, or reduce the suction flow rate, or take other steps to ensure that groundwater is not sucked up by the suction well.

b) Measure the suction pressure in the suction well over time and the generated pressure at each observation well. The generated pressure in the observation wells shall be determined by adding normal pressure to the pressure differential measured by the pressure differential gauge.

c) Conclude the test at the point at which the generated pressure at each observation well is constant.

6 Analysis of results

Calculate the suction flow rate of the gas in the soil under steady-state conditions as determined by the constant flow rate suction test, and the air permeability of the gas in the soil as determined by the suction pressure in the suction well and the generated pressure in each observation well, using the following equation. The following equation can be applied only when the ground surface does not permit air to pass through easily and the suction well screen is placed at a depth that encompasses the entire target vadose zone.

$$k_a = \frac{2.3 Q_P P_0 \mu}{\pi b (P_i^2 - P_1^2)} \log\left(\frac{r_i}{r_1}\right)$$

where

- k_a : Air permeability of gas in soil (m^2)
- Q_P : Suction flow rate of gas in soil during constant flow rate suction (m^3/s)
- P_0 : Normal pressure (101.325 kPa)
- P_1 : Generated pressure at observation well nearest to suction well (kPa)
- P_i : Generated pressure at observation well i (kPa) (i=2, ..., n)
- r_1 : Distance from center of suction well to center of observation well nearest to suction well (m)
- r_i : Distance from center of suction well to center of observation well i (m)
- μ : Coefficient of viscosity of air (1.82×10^{-8} kPa·s at normal temperature 20 °C and normal pressure 101.325 kPa)
- b : Thickness of target vadose zone (m)

Use the following procedure to determine the air permeability of the gas in the soil.

a) Calculate $P_i^2 - P_1^2$ from the generated pressure at the observation well nearest to the suction well and the generated pressure at each observation well.

b) Calculate $\log(r_i/r_1)$ from the distance from the center of the suction well to the observation well closest to the suction well, and the distance from the suction well to the observation wells.

c) Create a graph with $P_i^2 - P_1^2$ as the vertical axis and $\log(r_i/r_1)$ as the horizontal axis.

d) Determine the gradient E of an approximation straight line for $P_i^2 - P_1^2$ and $\log(r_i/r_1)$, as shown in Fig. A.2 in Annex A.

e) Determine the air permeability of the gas in the soil from gradient E and the suction flow rate Q_P of the gas in the soil during constant flow rate suction, etc. using the following equation.

$$k_a = \frac{2.3 Q_P P_0 \mu}{\pi b E}$$

Note 1: The presence or absence of air intruding from the surface of the ground shall be determined by the covering status from pavement and so on. In the event that air has intruded from the ground surface, a calculation equation that considers vertical flow as shown in the commentary is used.

Note 2: Fig. A.2 in Annex A shows an example of the analysis of the results of the air permeability test at the site.

7 Reporting

The following items shall be reported.

a) Suction well and observation well number and location

b) Height of top of suction well and observation wells

c) Structural diagrams of suction well and observation wells

d) Target ground soil classifications and distribution

e) Air pressure and temperature

f) Pressure measurement apparatus and method

g) Suction flow rate measurement apparatus and method

h) Test date and time

i) Results of constant flow rate suction test (changes over time in suction flow rate and suction pressure in gas in soil, relationship between generated pressure and distance)

j) Air permeability of gas in soil in target ground

k) If the method used deviates in any way from this standard, give details of the method used.

l) Other reportable matters

Annex A
(Reference)

Example of placement of air permeability test equipment at site and diagrams showing analysis of results

A.1 Example of placement of air permeability test equipment at site

Fig. A.1 shows an example of the placement of the air permeability test equipment at the site.

A.2 Example of analysis of results

Fig. A.2 shows an example of the analysis of results.

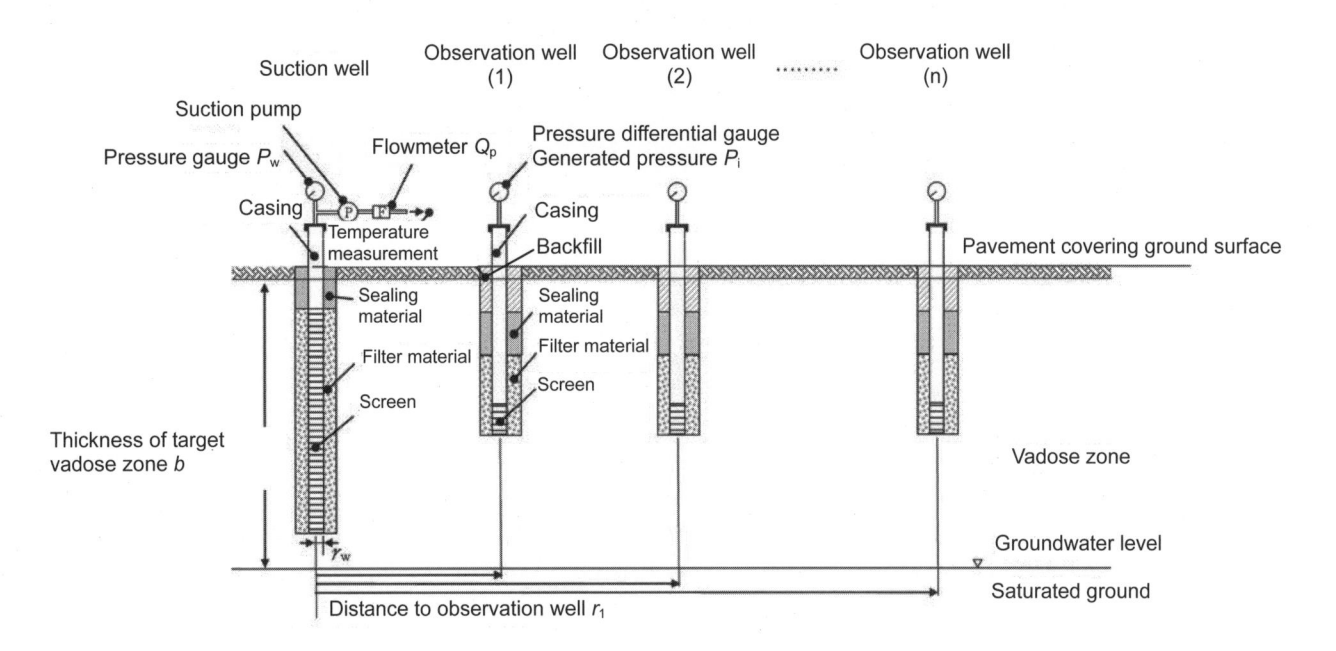

Fig. A-1 Example of placement of air permeability test equipment at site

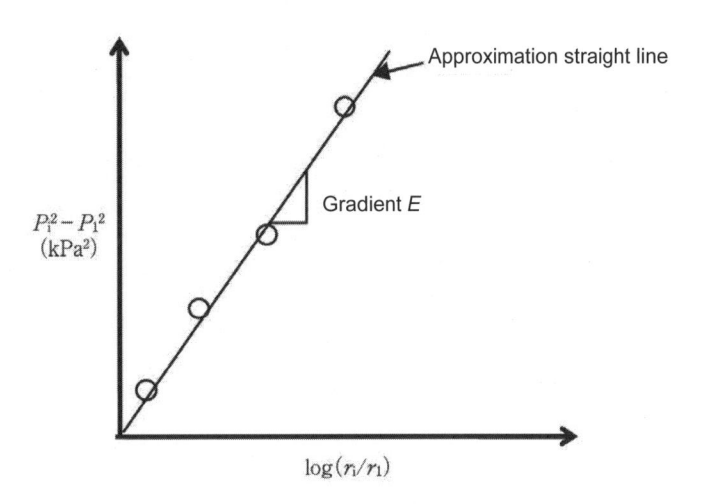

Fig. A-2 Example of analysis of results

List of Sponsors

EAST JAPAN RAILWAY COMPANY
KAJIMA CORPORATION
SHIMIZU CORPORATION
TAISEI CORPORATION
TOKYO SOIL RESEARCH CO., LTD.
FUDO TETRA CORPORATION
RAITO KOGYO CO., LTD.
CHIBA ENGINEERING CORP.
CHUO KAIHATSU CORPORATION
DIA CONSULTANTS CO., LTD.
GEO-LABO CHUBU
GIKEN LTD.
JAPAN PORT CONSULTANTS, LTD.
KINJO RUBBER CO., LTD.*
KISO-JIBAN CONSULTANTS CO., LTD.
NAKANIHON ENGINEERING CONSULTANTS CO., LTD.
NEWJEC INC.
NIPPON KOEI CO., LTD.
NITTOC CONSTRUCTION CO., LTD.
OBAYASHI CORPORATION
OKUMURA CORPORATION
OYO CORPORATION
SUMITOMO MITSUI CONSTRUCTION CO., LTD.
TOKYO ELECTRIC POWER SERVICES CO., LTD. (TEPSCO)*
TOKYU CONSTRUCTION
TOYO CONSTRUCTION CO., LTD.
YBM CO., LTD.
ASANO TAISEI KISO ENGINEERING CO., LTD.
CHEMICAL GROUTING CO., LTD.
CTI ENGINEERING CO., LTD.
INTEGRATED GEOTECHNOLOGY INSTITUTE LIMITED
KAWASAKI GEOLOGICAL ENGINEERING CO., LTD.
KUMAGAI GUMI CO., LTD.
MAEDA CORPORATION
NISHIMATSU CONSTRUCTION CO., LTD.
ONODA CHEMICO CO., LTD.
SANSHIN CORPORATION
TEKKEN CORPORATION
TENOX CORPORATION
KANSAI GEO AND ENVIRONMENT RESEARCH CENTER
HOKKAIDO SOIL RESEARCH CO-OPERATION
CHODAI CO., LTD.*
CHUDEN ENGINEERING CONSULTANTS

List of Sponsors (continued)

CHUO FUKKEN CONSULTANTS CO., LTD.
COASTAL DEVELOPMENT INSTITUTE OF TECHNOLOGY
EIGHT-JAPAN ENGINEERING CONSULTANTS INC.
ETERNAL PRESERVE*
FUJITA CORPORATION
FUKKEN CO., LTD.
HAZAMA ANDO CORPORATION
HOJUN CO., LTD.
HOKUKOKU CHISUI CO., LTD.
JAPAN CONSERVATION ENGINEERS & CO., LTD.
MARUI & CO., LTD.
NIKKEN SEKKEI CIVIL ENGINEERING LTD.
NIPPON ENGINEERING CONSULTANTS CO., LTD.*
NiX CO., LTD.
ORIENTAL CONSULTANTS CO., LTD.*
ORIENTAL CONSULTANTS GLOBAL CO., LTD.*
THE OVERSEAS COASTAL AREA DEVELOPMENT INSTITUTE OF JAPAN (OCDI)
PACIFIC CONSULTANTS CO., LTD.
PENTA-OCEAN CONSTRUCTION CO., LTD.
SATO KOGYO CO., LTD.
SERVICE CENTER OF PORT ENGINEERING (SCOPE)
SOIL AND ROCK ENGINEERING CO., LTD.
TAKENAKA CIVIL ENGINEERING & CONSTRUCTION CO., LTD.
TAKENAKA CORPORATION
TOA CORPORATION
TOA GROUT CO., LTD.*
WESCO CO., LTD.
* No AD.

SHINKANSEN
SERIES E5 SERIES E6 SERIES E7

JR-EAST

To explain the mysteries of the universe,
We achieved digging tunnels at tremendous speeds.

Two tunnels stretch straight ahead for 3000 meters.

In Kamioka town, Hida City, Gifu Prefecture, home of the "Super-Kamiokande" observatory,

the tunnels had been constructed for the purpose of setting up a large-scale cryogenic gravitational wave telescope.

The KAGRA is a device used to measure "gravitational waves",

planned by the Institute for Cosmic Ray Research at the University of Tokyo.

Gravitational waves are physical phenomena of ripples in the curvature of space-time.

To respond to the hopes of researchers to begin observations as soon as possible

so they can become the first in the world to detect gravitational waves,

Kajima Corporation dug the tunnels at an unprecedented maximum speed of 359 meters a month.

Kajima Corporation will continue to respond to the needs

of many people through its technology.

KAJIMA CORPORATION

Strength from imagination.

French novelist Jules Verne once said, 'Anything that one man can imagine, another man can make real.' Things we now take for granted — like cell phones and rockets — seem straight out of the world Verne imagined over 100 years ago. Imagination opens up real possibilities. Today, the impact of each new structure we build on earth comes under close scrutiny. What we must build next is a healthy relationship between humanity and our planet. That means imagining life's needs one hundred, two hundred, or more years from now. In doing so we are transforming the technical challenges of today into the common-sense solutions of tomorrow.

Kajima pioneers new frontiers — creating architecture and infrastructure for centuries to come.

鹿島
KAJIMA CORPORATION

The Future Begins with a Dream

Our dream is to make the world more comfortable, secure and sustainable. In developing each of our technologies, we continually envision a future in which we "live in harmony with nature."

Today's Work, Tomorrow's Heritage

SHIMIZU CORPORATION
SHMZ

The Environmental Island GREEN FLOAT

A marine city floating on the waters of the Pacific. By combining nature's boundless stores of solar, wind and marine energy with the latest technological innovations, we can ensure a self-sustaining supply of energy and food that goes beyond carbon neutral into carbon negative. This "botanical city" provides a path towards sustainable co-existence in which people and nature can live in comfort and harmony.

Shimizu Dream | GO!

A&S Soil Improvement Method (Absorption and Subsidence Soil Improvement Method)

Summary

This method is able to improve soil property in a short term by 3 points.
① Suctioning pore water by Super Well Point Method(SWP)
② Improve permeability by inspiration and expiration in the ground
③ Promote consolidation settlement by loading fill

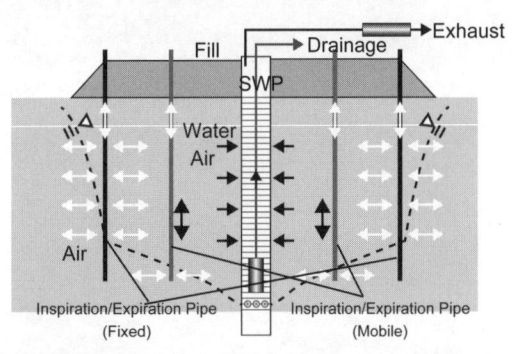

Diagrammatical view: A&S Soil Improvement Method

Feature

① Subsidence convergence in only 3months
② The cost will be half or less than deep mixing method
③ The applicable depth is around 50m (Experience:30m)
④ It controls residual settlement surely
⑤ Compact construction facilities
⑥ Only the pipe of the SWP is left

Cunstruction

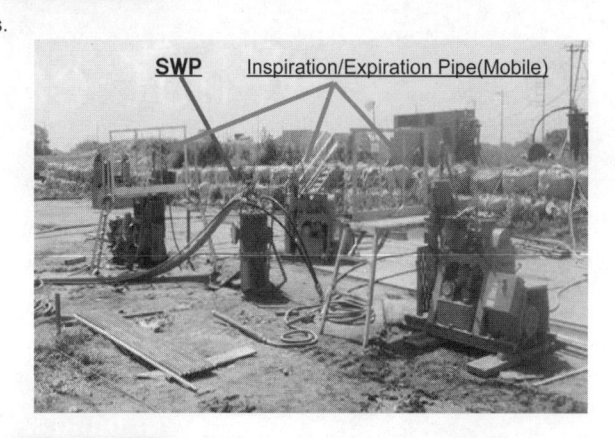

Result

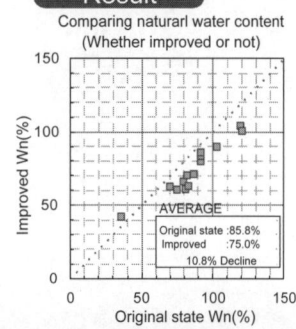

Comparing naturarl water content (Whether improved or not)
AVERAGE
Original state : 85.8%
Improved : 75.0%
10.8% Decline

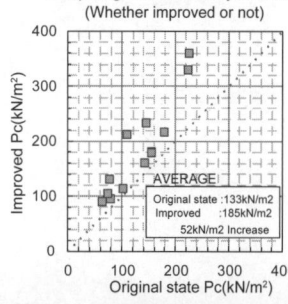

Comparing Consoilidation yield stress (Whether improved or not)
AVERAGE
Original state : 133kN/m2
Improved : 185kN/m2
52kN/m2 Increase

Gravel Support Method

Summary

The gravel support method reduces liquefaction-induced damage of small structures or outdoor facilities by improving only the ground surface with highly permeable gravel layer.

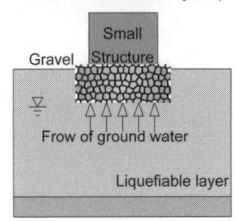

newly-constructed structures (Gravel layer)

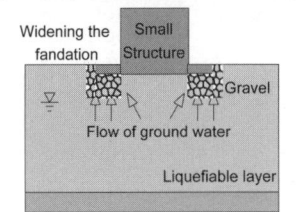

pre-existing structures (Gravel layer+ Widening the foundation)

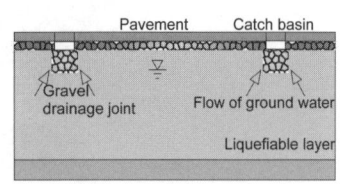

outdoor facilities (Gravel drainage joint)

Applicability

(1) Structures which meets all four conditions below
　① Weight : less than 5t/m^2
　② Short side length : less than 15m
　③ Thickness of the foundation (pre-existing structures only) : larger than 50cm
　④ No habitable room
(2) Outdoor facilities

Advantage

① Liquefaction-induced damages can be largely reduced by improving only the surface layer
② Can significantly reduce the cost in compare with other existing methods
③ Is application to wide variety of construction conditions
④ Can be constructed without stopping the operation of facilities

Effect of the measures

Verification of the effect of the measures by centrifugal model tests

No measures

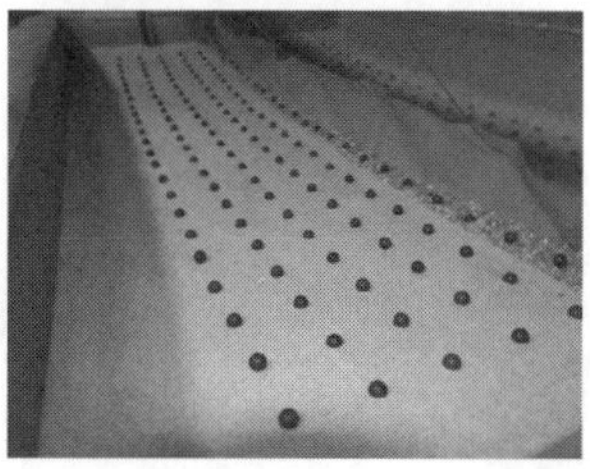

Gravel Support Method

Today's Work,
Tomorrow's Heritage

SHIMIZU CORPORATION
SHMZ

What can be done?

At Taisei, we will continue to create

a vibrant environment for all members of society

in order to build a global community filled

with hopes and dreams.

TAISEI CORPORATION
For a Lively World

For a Lively World

A city of the future envisioned by Taisei's technology

"To create a vibrant environment for all members of society."
Adhering to this Taisei Group Ideal, we have created safe and attractive
spaces as well as work and living environments of exceptional value with
the aim of providing products and services that are in harmony with nature.
For a Lively World — embracing this Group Slogan, we enhance and pass on
the technologies we have further refined to the next generation.
We also work to support the creation of a wholesome global society that will
benefit all of humanity and brimming with hopes and dreams.

Lively Earth, Lively People.   Taisei's quest for the future.

TAISEI CORPORATION

TOKYO SOIL RESEARCH

Over 50-year History

Tokyo Soil Research provides a comprehensive range of engineering and consulting services such as geotechnical, geological, geophysical and geoenvironmental site investigation and laboratory tests, planning and design of environmental protection and disaster prevention for construction projects, site-specific ground motion prediction, seismic diagnosis and structural retrofit design, maintenance, and rehabilitation, using state-of-the-art, cutting-edge technology and expertise.

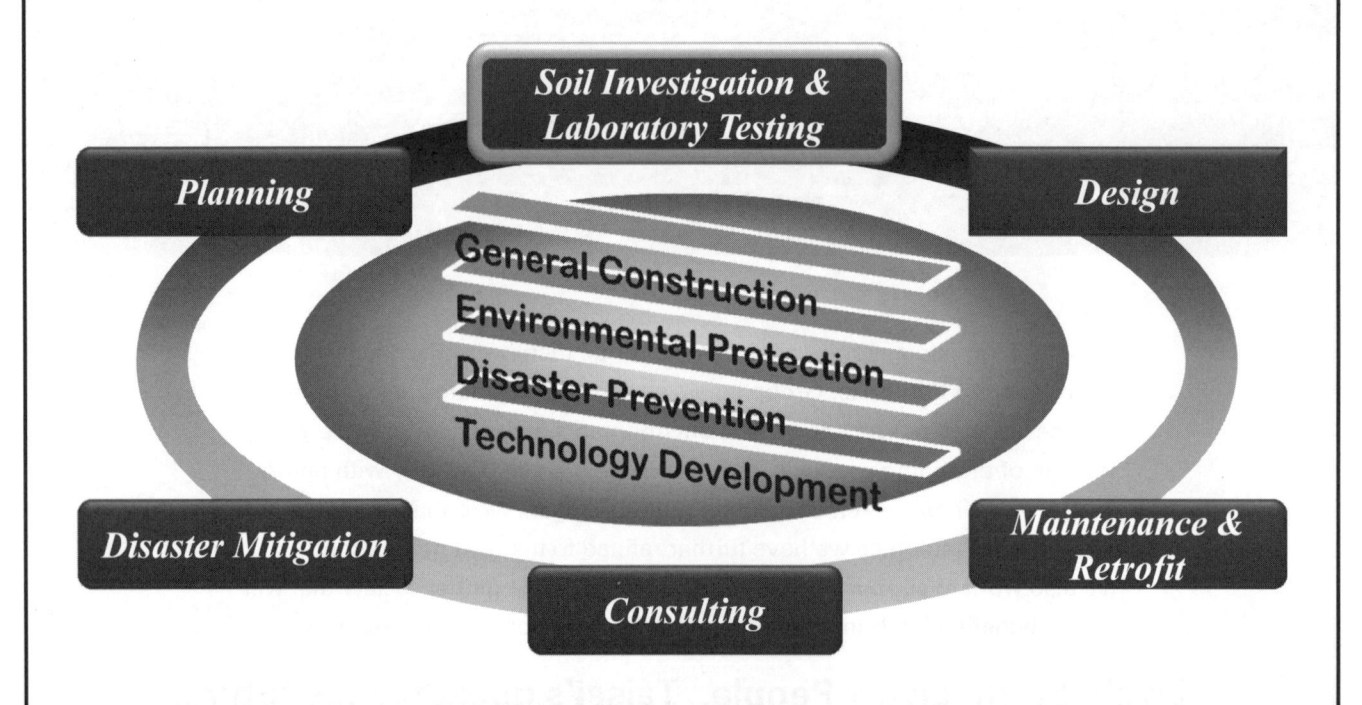

- Soil Investigation & Laboratory Testing
- Planning
- Design
- Disaster Mitigation
- Maintenance & Retrofit
- Consulting

General Construction
Environmental Protection
Disaster Prevention
Technology Development

Comprehensive and Innovative Engineering Solutions to Your Construction Needs

TOKYO SOIL RESEARCH CO., LTD

Head Office
2-11-16 Higashigaoka, Meguro-ku, Tokyo 152 - 0021, Japan
TEL + 81-3-3410-7221 / FAX + 81-3-3418-0127 / E-mail : info@tokyosoil.co.jp
URL: http://www.tokyosoil.co.jp

Key Expertise and Services

General Construction

Long-span bridge
(Aichi - pref)

- Soil Investigation and Geological Survey using Various Types of In-Situ Tests
- Geological Reconnaissance
- Geophysical Exploration and Geophysical Logging
- Ground Water Investigation
- Hydrological Survey
- Laboratory Tests to Determine Engineering Properties of Soil and Rock
- Strength Tests on Rock Mass for Engineering Products
- Ground Vibration Measurement
- Monitoring of Deformation During and After Construction
- Analysis of Stress Deformation
- Earthquake Response Analysis
- Soft Ground Analysis
- Analysis of Groundwater Flow
- Wide Range of Material Tests

Environmental Protection

Historical Building
(Nara - pref)

- Investigation of Soil and Ground Water Contamination
- Investigation of Soil Gas
- Chemical Analysis of Soil and Groundwater
- Measurement of Groundwater Flow Direction and Flow Velocity
- Proposal on Disposal of Contaminated Soil and Groundwater
- Monitoring of Contaminated Sites During Treatment
- Examination and Conservation Survey for Ruins, Buried Cultural Properties and Historic Buildings

Disaster Prevention

Earthquake-induced damage during 1995 Kobe Earthquake

- Compilation of Geotechnical, Geological and Geophysical Database
- Investigation and Testing of Structures
- Evaluation of Seismic Performance of Existing Building
- Seismic Retrofit Design of Existing Building
- Creation of Simulated Earthquake Waveform
- Dynamic Analysis of Ground and Structures
- Disaster Prevention Design of Slopes

Technology Development

IT-system

- Sampling of Sand and Gravel by In-Situ Freezing
- Sampling of Contaminated Soil
- Microtremor Array Observation for S-Wave Velocity Profiling
- Pile Integrity Test (IT-system)
- Measurement of Vertical and Horizontal Permeability using Single Borehole

Comprehensive and Innovative Engineering Solutions to Your Construction Needs

TOKYO SOIL RESEARCH CO., LTD

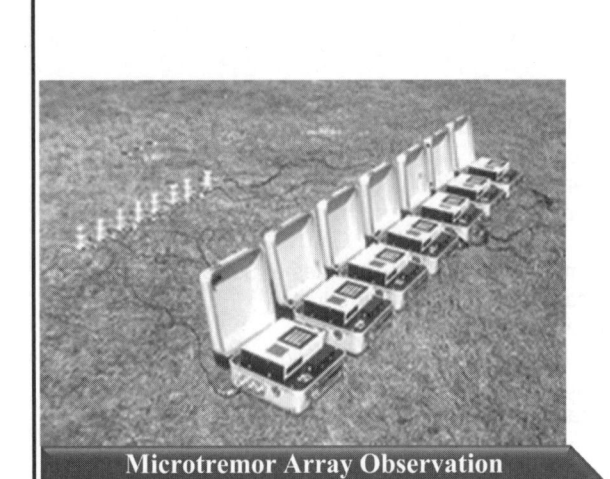

Microtremor Array Observation for S-Wave Velocity Profiling

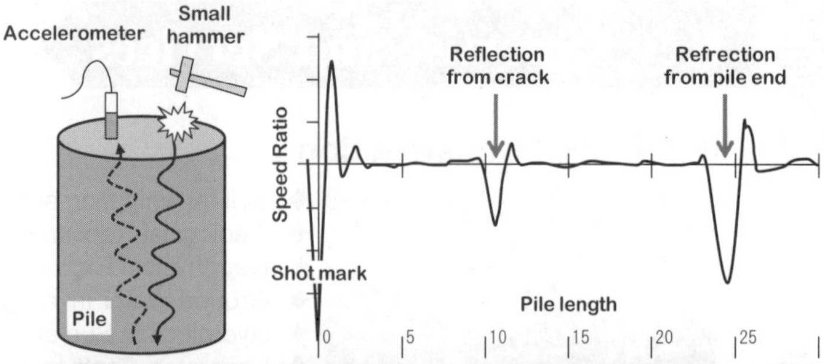

Pile Integrity Test < IT - System >

Our wide scope of engineering and consulting services ranging from field and laboratory testing, geotechnical and foundation engineering, to ground motion prediction, seismic diagnosis and retrofit design of structures to suit your construction needs.

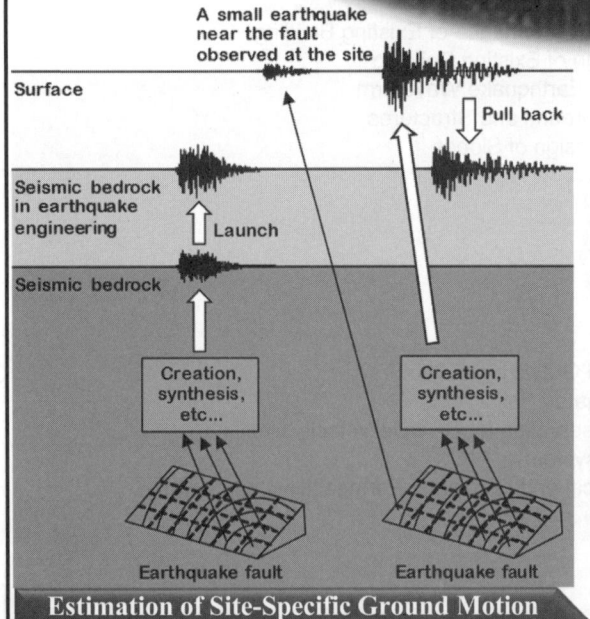

Estimation of Site-Specific Ground Motion for Structural Design

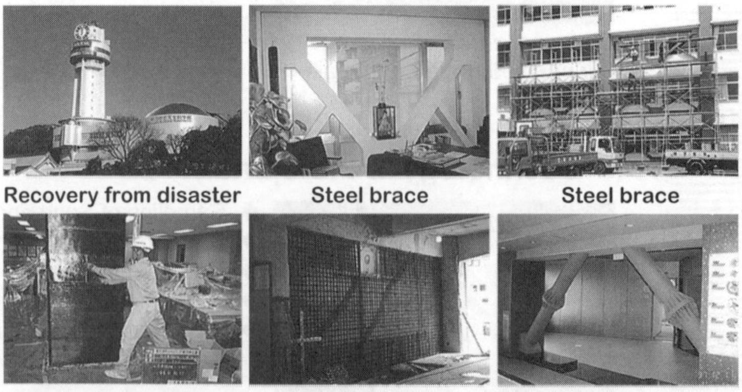

Seismic Diagnosis of Existing Building and Retrofit Design

- Recovery from disaster
- Steel brace
- Steel brace
- Winding carbon fiber
- Resistant wall
- Steel pipe brace

Comprehensive and Innovative Engineering Solutions to Your Construction Needs

TOKYO SOIL RESEARCH CO., LTD

Sampling of Sand and Gravel by In-situ Freezing

High quality undisturbed samples from sandy and gravelly layers obtainable with state-of-art in-situ freezing and sampling of frozen soils by core boring, which are considered difficult to be preserved by any conventional methods.

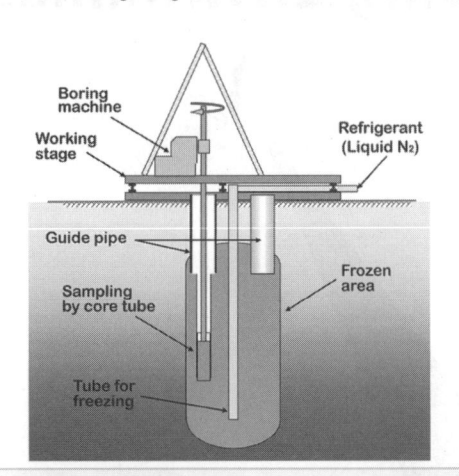

Samples of Gravel

Samples of Sand

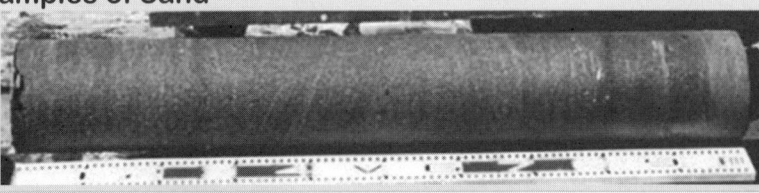

Core Tube Tip

Small to Large Cyclic Triaxial Test

Triaxial shear testing with specimen sizes ranging from the smallest diameter of 50 mm and height of 100 mm, to the largest diameter and height of 300 mm and 600 mm respectively.

Torsional Shear Test

Comprehensive and Innovative Engineering Solutions to Your Construction Needs

TOKYO SOIL RESEARCH CO., LTD

Setting our course towards…
A Safe and Secure Living Environment

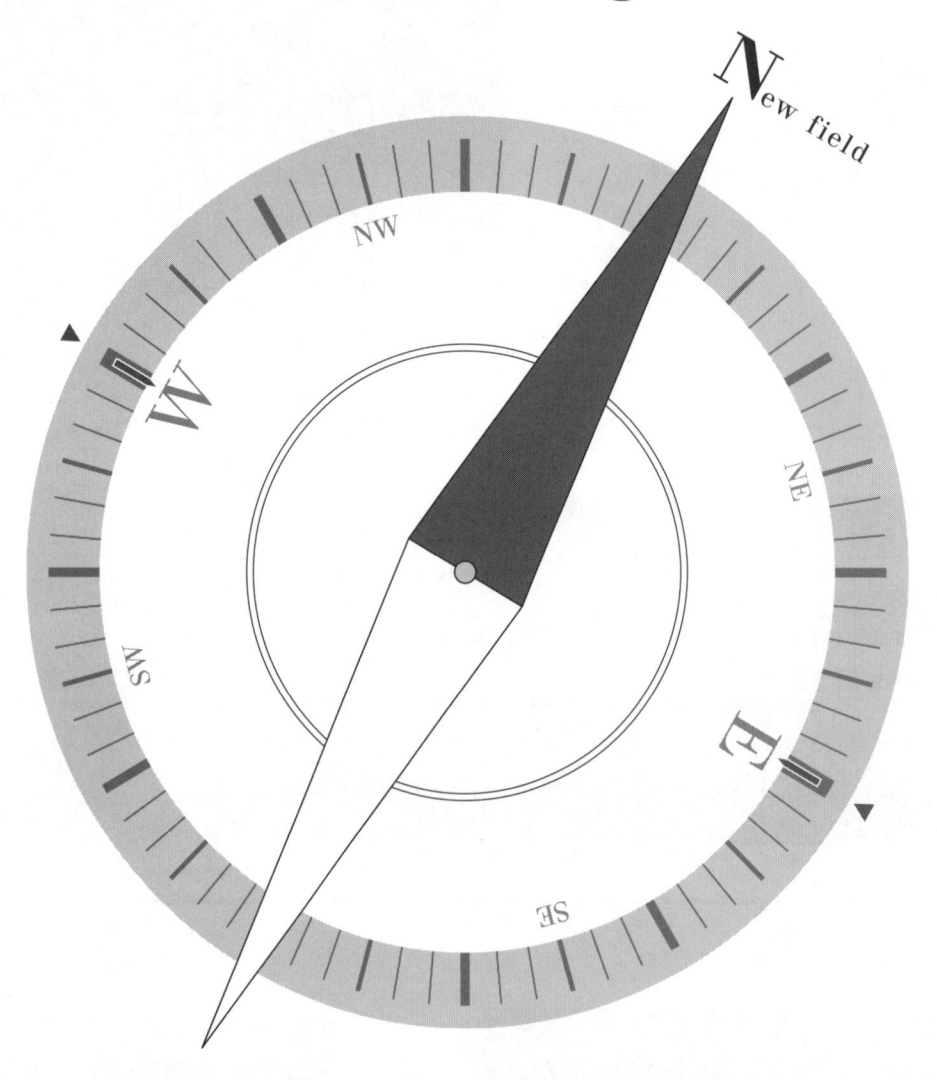

We at Fudo Tetra Group use our own technologies and innovative concepts to help build a secure and comfortable living environment.

Our new fields extend our reach from the sea bed to mountain tops, to make our land more resistant to natural disasters and to create a social environment for modern lifestyles that brings peace of mind.

Fudo Tetra Corporation

7-2 Nihombashi Koami-cho, Chuo-ku, Tokyo 103-0016
Tel: +81 (0) 3-5644-8500

http://www.fudotetra.co.jp

Unique Technologies and Know-How Playing a Major Part in the Construction of Social Infrastructure

CIVIL ENGINEERING

With our wealth of technical expertise and experience we are building the social fabric of the future in the following fields:
- Roads, railways, dams and waterways, water supply and sewerage systems, energy infrastructure and other land-based civil engineering
- Port and airport installations, fishing facilities, coast defenses, man-made islands and other coastal and offshore civil engineering

Main technologies:
Mountain tunneling; shield tunnel boring; dam construction; earthworks; offshore engineering; earthquake resistance strengthening and replacement;
waste disposal plant construction; dismantling of incineration facilities; water treatment; soil decontamination

SOIL IMPROVEMENT

In 1956 we were first in the world to develop the sand compaction pile method (SCP), which is so effective for mitigation liquefaction during earthquake. We continue to build on our R&D record, and as experts in soil improvement we have a comprehensive range of design and implementation technologies and a portfolio of project implementation.

Main technologies:
Compaction, de-watering, drainage, solidification and water cut-off methods; lightweight soil; anti-vibration technologies

ENVIRONMENT-USE BLOCKS

We lease forms for concrete blocks including tetrapods and other wave-dissipating blocks, and for waterfronts we offer various technologies and design services, and also products designed to protect coastal aspects and ecosystems.

Main products:
Shaped concrete blocks of tetrapod, tetraneo, pelmex and other designs; filtering units; ion culture components; etc.

Fudo Tetra Corporation
FUDOTETRA

**High Reliability and Quality
with specialized technology and abundant ideas**

Raito — GEOTECHNICAL SPECIALIST
E-mail. overseas@raito.co.jp

Japanese Geotechnical Society

4-38-2 Sengoku, Bunkyo-ku, Tokyo, 112-0011, Japan

Tel: +81-3-3946-8677   Fax: +81-3-3946-8678   E-mail: jgs@jiban.or.jp

Aiming at coexistence with the earth
Overall engineering on earth

【Site Investigation】
We accumulate detailed and accurate geotechnical information through conscientious field investigation.

【Soil Test】
We characterize the physical and mechanical properties of the soil based on sophisticated sampling and testing technology.

【Foundation Analysis】
We perform accurate analytical estimation of the ground behavior by using the results of site investigation and laboratory soil test.

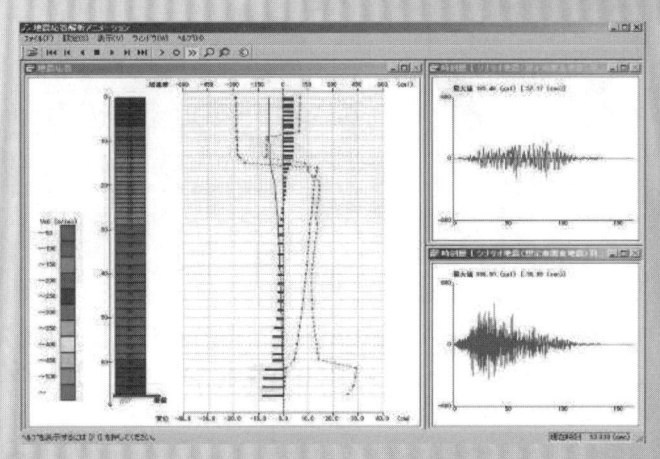

Chiba Engineering Corp.

1-30-5, Makuharihongo, Hanamigawa-ku, Chiba 262-0033, Japan
TEL. +81-43-275-2311   FAX. +81-43-275-4711
http://www.chiba-eng.co.jp

Early Warning System for Slope Failure and Landslide

Slope Failure Monitoring Sensor 【KANTARO】
斜面崩壊感知センサー 感太郎
※Join development with Univ. of Tokyo

KANTARO is a sensor which catches the changes of tilt-angle and water content inside the superficial layer of the slope. It can saves data to SD card and transfers data out using wireless communication.

- ☑ East installation
- ☑ Low cost of equipment and installation
- ☑ Power saving (Four C-size-batteries work about one year)
- ☑ Waterproof (JIS standard IPX7)
- ☑ High precision (Sensor measurement accuracy: 0.017°)
- ☑ Established method of evaluating risk and warnings for a dangerous slope:
  - Tilt rate ≧ 0.03° per 3 hours (= 0.01° / hr.) → Precaution
  - Tilt rate ≧ 0.1° per hour → Warning
- ☑ Construct a stable early warning system using multi-point measurement

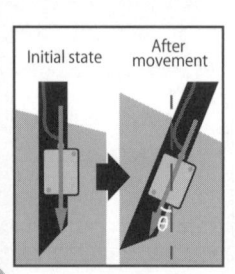

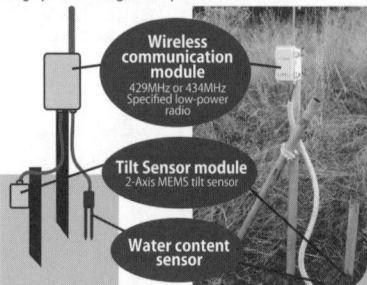

Initial state / After movement

Wireless communication module
429MHz or 434MHz Specified low-power radio

Tilt Sensor module
2-Axis MEMS tilt sensor

Water content sensor

Multistage Inclinometer 【K-TA】
多段式傾斜センサー K太
※Join development with Univ. of Tokyo

An easy installation multistage inclinometers that can gather the displacement value of each segment based on combining tilt angle data with direction data (measured by embedded digital compass).

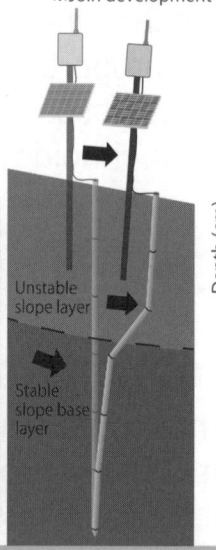

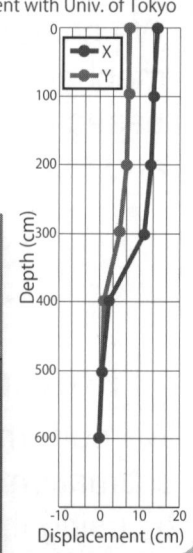

Unstable slope layer / Stable slope base layer

Automatic Internet Full Duplex Remote Monitoring System 【KANSOKUO】
双方向遠隔自動監視システム 観測王

Sensors as Tilt sensor, Surveillance camera, Water gauge, Seismometer, Rain gauge...etc.

Data logger → Server → Web (Monitoring data at Internet ; Landslide disaster alert mail) / Mobile device / User PC

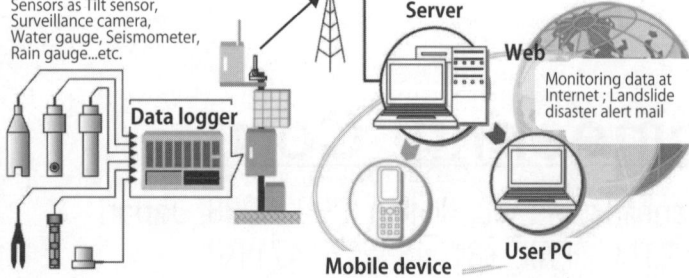

「KANSOKUO」 is our remote monitoring system that transmits data of different types of sensors and control the local instrument remotely by full duplex communication. This system can be used to landslide and slope failure monitoring, debrisflow and rock fall monitoring and riverbank leakage monitoring.

Overseas Achievements

Country	Area	Year	Purpose
China	Three Gorges Dam, Wanzhou, Sichuan	2009	Land slide
	Tazhiping, Dujiangyan, Sichuan	2010	Land slide
	Wangjiagou, Dujiangyan, Sichuan	2013	Debris flow
	Hongmeichun, Dujiangyan, Sichuan	2013	Land slide
	Yingdongzigou, Dujiangyan, Sichuan	2013	Debris flow
	Gangou, Dujiangyan, Sichuan	2013	Debris flow
	Yuexi Baixiangyun, Liangshan, Sichuan	2015	Debris flow
	Tanjiazhui, Longquan, Sichuan	2015	Land slide
	Donglonggou, Songpan, Sichuan	2015	Debris flow
	Qinglongsi, Penzhou, Sichuan	2015	Land slide
	Liangjiashan, Chongzhou, Sichuan	2015	Land slide & Debris flow
	Hetaogou, Maoxian, Sichuan	2016	Debris flow
	Yushigou, Penzhou, Sichuan	2016	Debris flow
Taiwan	Yanchao, Kaohsiung	2015	Land slide
	Meishan, Chiayi	2015	Land slide
	Alishan, Chiayi	2015	Land slide
	Chaishan, Kaohsiung	2016	Land slide
	Zhongpu, Chiayi	2017	Land slide
	Zhuqi, Chiayi	2017	Land slide
Sri Lanka	Uva, Badulla	2016	Land slide
Australia	Maleny, Queensland	2016	Land slide
India	Kalimpong, West Bengal	2017	Land slide

Specific Laboratory Tests for Soil and Rock

HYBRID type triaxial equipment

In coastal area of Japan, ground liquefaction cause by earthquake has brought great damages. As one method of determining reason of disasters caused by liquefaction, we developed HYBRID type triaxial apparatus that can execute dynamic loading test or static loading test, in addition to general soil laboratory test.

Results of Random Wave Loading Test

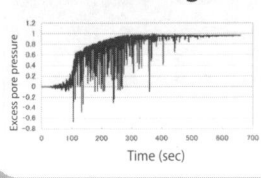

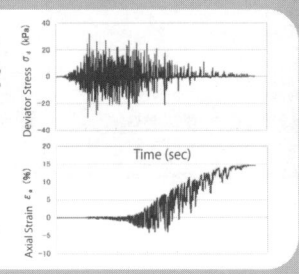

Results of Cyclic Loading Test

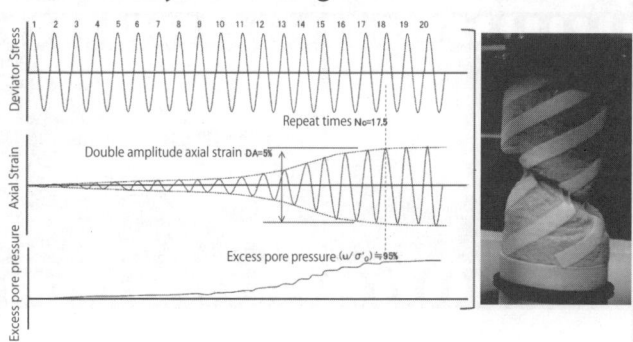

An example of a liquefaction test results(left) and specimen after liquefaction(right).

Vendor Element Test

There is a need for rapid liquefaction countermeasures for irrigation pond. The elucidation of the behavior during an earthquake, a high precision analysis is required. Our developed test equipment can meet the request, reproduce the damage mechanism and improve the special analysis technique (SERID).

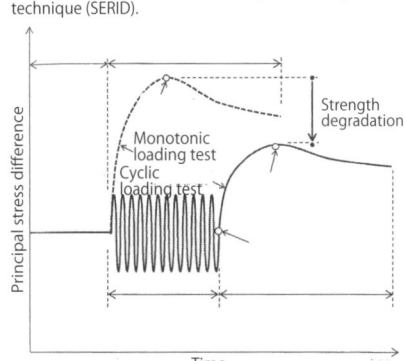

Destruction situation of irrigation pond

Vendor element (Disk Type)→

中央開発株式会社
Chuo Kaihatsu Corporation

Technical Support: Technology Center / Disaster Prevention Division
Tel:+81-3-3208-5252 / Fax:+81-3-3232-3625
3-13-5 Nishi-Waseda, Shinjuku-Ku, Tokyo 169-8612, Japan
E-mail: ckc_post@ckcnet.co.jp / http://www.ckcnet.co.jp

DIA CONSULTANTS

http://www.diaconsult.jp/

Thinking hard about the human being in harmony with the earth.

Our Vision

Envioronments

The increasing of carbon dioxide is the major cause of global warming problem. Our company has participated and contributed to various projects for reducing the carbon dioxide emission such as carbon capture storage project in recent years.

Natural Disasters

Japan is the top prone to earthquake country in the world. The extensive damages are caused by the large-scale of earthquake and tsunami. In addition, the abnormal weather causes the frequently river flooding. Our company performs various investigations and analysis to support the development of disaster prevention plan.

Energy Resources

In recently years, the depletion of fossil fuels has become a major problem. Our company has participated and contributed to the national oil and liquefied propane projects as the energy security program.

DIA CONSULTANTS

Head Office
Add: 1-7-4, Iwamotocho, chiyodaku, Tokyo 101-0032, JAPAN
Tel:+813 5835 1711, Fax:+813 5835 1720

One Stop Service

Our company provides one stop service for a client starting from the investigation until the maintenance service after completion of a construction.

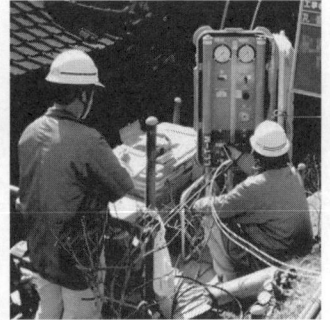

Subsurface Exploration

Civil Engineering Structural Design

Structural Maintenance

Numerical Analysis

Technology

Soft ground investigation

The seismic earthquake shall be considered to the construction design over soft ground area such as landfill and reclaimed land.
Our company utilizes the latest technology of soft ground investigation and proposes the best solution and method of a construction work.
In addition, we are developing the advanced subsurface investigation method closely with industrial, academic and government sector.

Active fault investigation

The construction of civil engineering structural shall be considered the effect of the earthquake.
The earthquake caused by the fault movement is increasing and can be roughly predicted the damage scale by examining the active faults in a vicinity of construction site.
Our company offers a high degree of reliable data by the advanced investigation technology over half a century.

DIA CONSULTANTS

Head Office
Add: 1-7-4,Iwamotocho,chiyodaku,Tokyo 101-0032, JAPAN
Tel:+813 5835 1711,Fax:+813 5835 1720

For a better society

Japan is located in one of the most active seismic zones in the world. After the 2011 Great East Japan Earthquake, our society seeks safety and people are hoping to have a more stable life.
In this age, Geo-Labo Chubu is contributing to developments for a better society with laboratory tests of geomaterials.

Nagoya Castle and buildings

Value creation
We pursue improvments of customer satisfaction in every aspect and are always creating new values.

Our mission

Contribution to society
We aim for the realization of a safe and secure society for all via laboratory tests of geomaterials.

Technological strength
We offer information of reliable geomaterials with ample facilities and skills that meet the demands of the present age.

Main test facilites

- We are specialists in laboratory tests of geomaterials -

Triaxial compression test

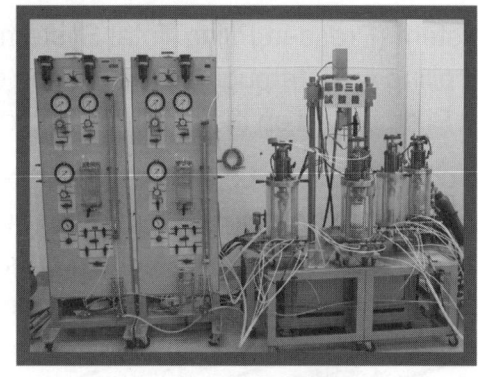

Cyclic triaxial test

Incremental loading consolidation test

Large scale triaxial test

■ Tests for physical properties of soils
Density test of soil particle / Water content test / Grain size analysis / Sedimentation analysis / Liquid limit test / Plastic limit test / Wet density test / Ignition loss test / pH test

■ Tests for mechanical properties of soils
Incremental loading consolidation test / Constant strain rate consolidation test / Unconfined compression test / Triaxial compression test / Liquefaction test / Cyclic triaxial test / Compaction test / California bearing ratio test / Cone index test / Permeability test / Large scale triaxial test

■ Tests for physical and mechanical properties of rocks
Ultrasonic pulse test / Slaking test / Permeability test for rocks / Point load test / Unconfined compression test for rocks / Swelling test / Water absorption test   etc.

Geo-Labo Chubu
Chubu Soil Reseach Cooperative Association

Address: Midorigaoka 804, Moriyama-ku, Nagoya, Aichi, Japan
Post code : 463-0009 / Telephone: +81-52-758-1500
Email: info@geolabo-chubu.com
*Please contact us the above Email for any questions.
http://www.geolabo-chubu.com/

Geo-Labo Chubu since 1979

Nagoya, Japan

Comprehensive control of construction and quality by the Press-in Method

PPT System

PPT System (Press-in Piling Total System) provides an objective quality control of piles or an efficient piling work through the estimation of subsurface information or the automatic operation of Silent Piler, based on the press-in piling data.

Ground conditions can be estimated and recorded throughout the piling area, and can be used for confirming the bearing stratum or for quality control as an observational method. Automatic operation can cope with the variation of ground conditions and reduces press-in time by 30% (compared with our existing automatic operation system).

EQUIPPED WITH PPTS
Press-in Piling Total System
AUTOMATIC OPERATION

Efficient piling work and objective quality control based on press-in piling data

//GIKEN

Countermeasure against flood, storm surge, earthquake and tsunami

IMPLANT LEVEE

Implant levee, consisting mainly of piles or sheet piles, is resilient to erosion due to flood, wave force due to storm surge or tsunami, and liquefaction due to earthquake, and maintains its function of disaster prevention without being failed completely.

Protect human lives, culture, history and property from natural disasters

Reinforcement of coastal levees (Kochi, Japan)

Renovation of retaining walls for roads (Kanagawa, Japan)

GIKEN LTD.

International Business Department
3948-1 Nunoshida, Kochi-shi, Kochi 781-5195, Japan
Tel. : +81-(0)88-846-2980   Fax : +81-(0)88-826-5288   Email : international@giken.com

www.giken.com

Global Network : Japan, USA, UK, Netherlands, Germany, Singapore, China, Australia

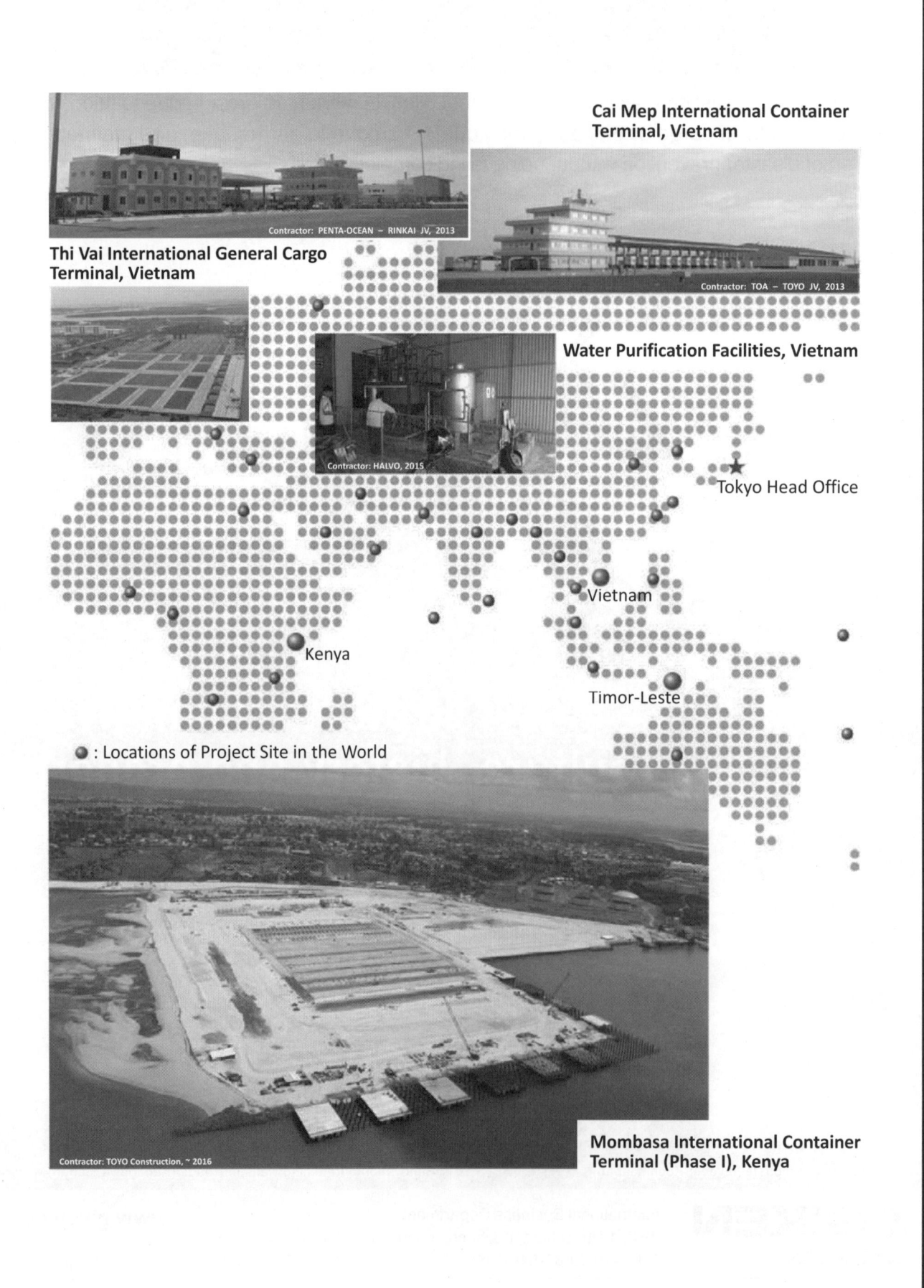

Cai Mep International Container Terminal, Vietnam
Contractor: TOA – TOYO JV, 2013

Thi Vai International General Cargo Terminal, Vietnam
Contractor: PENTA-OCEAN – RINKAI JV, 2013

Water Purification Facilities, Vietnam
Contractor: HALVO, 2015

Tokyo Head Office

Vietnam

Kenya

Timor-Leste

● : Locations of Project Site in the World

Mombasa International Container Terminal (Phase I), Kenya
Contractor: TOYO Construction, ~ 2016

Cai Mep International Container Terminal, Vietnam

Contractor: TOA – TOYO JV (Crane: IHI – MES JV), 2013

Oecusse Ferry and Cargo Terminal, Timor-Leste

Contractor: TOBISIMA Corporation, 2013

Japan [Head Office]   Tel: +81-3-5434-8163
Kenya [Mombasa Project Office]   Tel: +254-73-355-5050
Vietnam [Hanoi Representative Office]   Tel: +84-43-826-2930
Indonesia [Jakarta Representative Office]   Tel: +62-21-782-9334

The forerunner of port engineering in Japan, the most authoritative and prestigious private consulting engineering firm.

Japan Port Consultants, Ltd.

www.jportc.co.jp   TK Gotanda Bldg., 8-3-6, Nishi-Gotanda, Shinagawa-ku, Tokyo, 141-0031

Legend
Colored Area:
  Project Track Record
Headquarter:
  Tokyo
Overseas Office:
  Singapore, Kuala Lumpur,
  Jakarta, Hanoi, Manila

KISO-JIBAN CONSULTANTS CO., LTD.

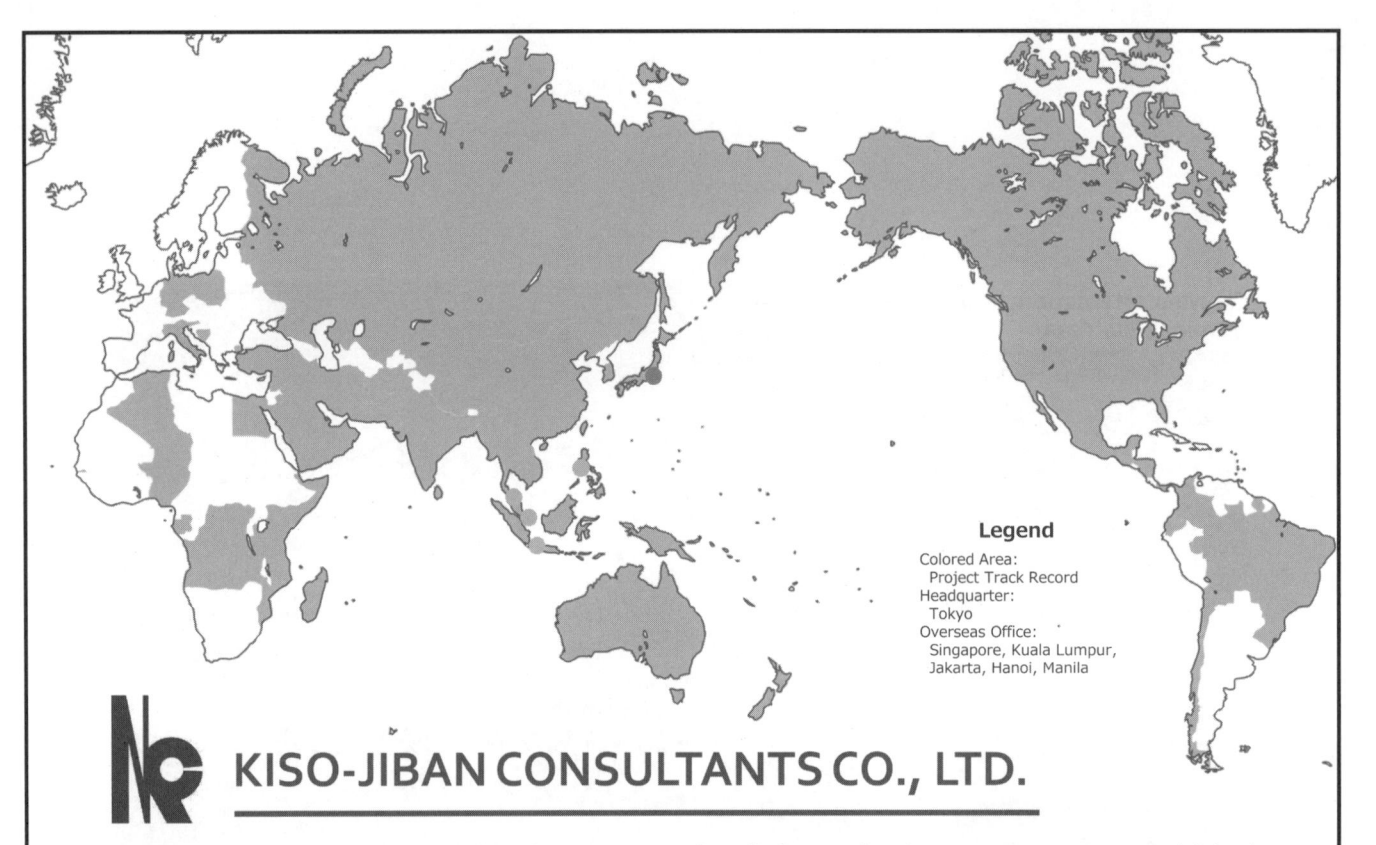

KISO-JIBAN is an engineering consulting firm experienced in all phases of civil engineering project. Established in 1953, we have participated in over 6000 overseas projects in more than 80 countries worldwide and have been leading the industry with our innovative technology as a pioneer.

Comprehensive Service Areas

Geological and Geotechnical Survey
Geotechnical Design and Analysis
Disaster Prevention and Management
GIS (Geographic Information Systems)

Soil and Rock Laboratory Tests
Instrumentation and Monitoring
Geophysical Exploration & Development
Distribution of Geosynthetics Products

Soil Investigation

Among existing sampling technologies, the most noteworthy technology should be Gel-Push Sampler which was developed by KISO-JIBAN. Gel-Push Samplers enable to obtain undisturbed soil samples on fragile soft ground.

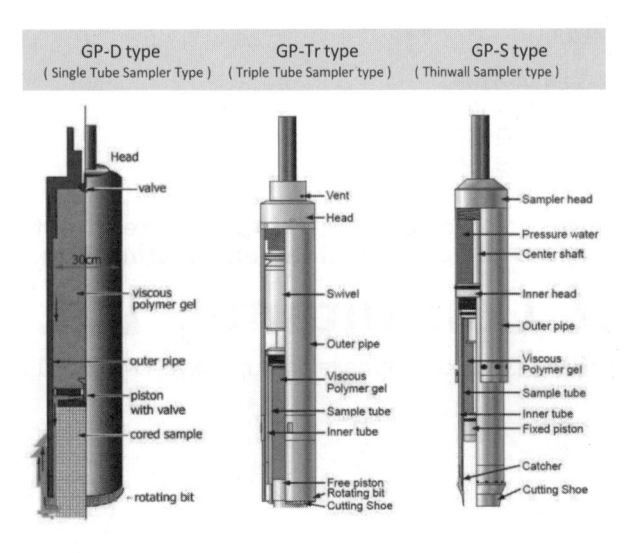

GP-D type	GP-Tr type	GP-S type
(Single Tube Sampler Type)	(Triple Tube Sampler type)	(Thinwall Sampler type)

Undisturbed Samples Obtained with GP Sampler

Overseas Offices

Tokyo

KISO-JIBAN Overseas Department
Kinshicho Prime Tower 12 Floor,
1-5-7 Kameido, Koto-ku,
Tokyo 136-8577, Japan
Tel: +81-3-6861-8886

Singapore

KISO-JIBAN Singapore Branch
KISO-JIBAN SINGAPORE PTE., LTD.
60, Kallang Pudding Road #02-00,
Tan Jin Chwee Ind. Bldg.,
Singapore 349320, Singapore
Tel: +65-67473233

Malaysia

KISO-JIBAN (MALAYSIA) SDN. BHD.
No.3, Jalan Keneri 17/D, Bandar Puchong Jaya,
47100 Puchong, Selangor Darul Ehsan, Malaysia
TEL: 603-80761377

Vietnam

KISO-JIBAN Representative Office
6th Floor, TID Building, 4 Lieu Giai Street,
Ba Dinh Dist., Hanoi, Vietnam
TEL: (84-4) 3232-1034

Indonesia

P.T. PONDASI KISOCON RAYA
Graha Sucofindo 14th Floor,
Jl. Raya Pasar Minggu Kav. 34,
Jakarta 12780, Indonesia
Phone: +62-21-7986663

Philippines

KISO-JIBAN Representative Office
Unit 2105, 88 Corporate Center,
Sedeno Street Corner Valero Street,
Salcedo Village, Makati City 1200, Philippines
TEL:+63-2-552-8238

Geophysical Survey & Analysis

KISO-JIBAN has developed and innovated a number of geophysical survey technologies as well as analysis software.

One of our notable geophysical survey method, 3D seismic refraction survey, enables the visualization of elastic wave velocity for underground structure.

For geotechnical analysis, we leverage our technical expertise and our experience in choosing method and forming judgment.

Being an international company, we are capable to work with all major international standards such as ASTM (American), BS (British), EN (European), DIN (German), and GOST (Russian).

As a pioneer, we are committed to keep listening to our clients and to move forward to the future.

3D Seismic Refraction Survey

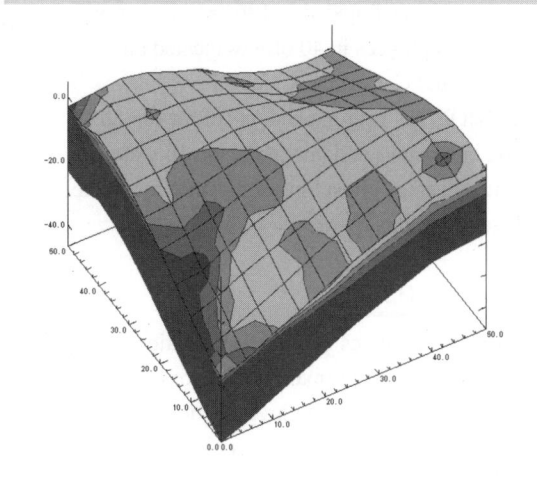

3D Terrestrial Fluid Analysis by "GETFLOWS"

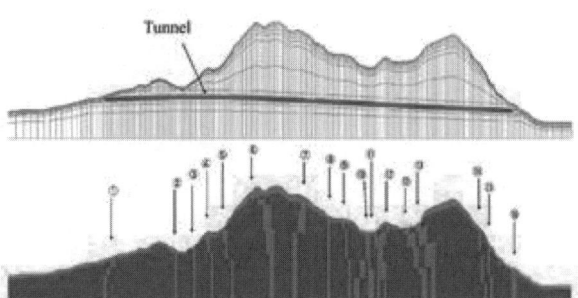

Match your desire with an advanced Technology

Contents of Main Engineering Services

ADVANCED 先進
- Asset Management
- PRE / PFI / PPP
- Network Sensor
- Oversea water business

LIFE 生活
- Water Works
- Drainage Works
- Recycle

EXCHANGE 交流
- Road
- Bridge / Tunnel
- Railways

HARMONY 調和
- Urban Planning / Park
- Environmental Assessment Analysis
- New Energy

SAFETY 安心
- River / Anti Tsunami
- Anti Flood
- Anti Earthquake
- Anti Disaster

Introduction

Nakanihon Engineering Consultants Co.,Ltd. (NAKANIHON) is an engineeringconsulting company headquartered in Nagoya, Japan.

NAKANIHON was founded in 1964 and has grown to about 350 employees in 40 offices located all around Japan and Cambodia.

NAKANIHON has been ranked one of the top 50 consulting companies and one of the top 10 water supply and sewerage consulting companies by Nikkei construction ranking.

Our Visions

NAKANIHON maintains strong client relationships through high quality performance. We provides a wide range of consultant tasks from investigation, planning, feasibility study, design, to site supervision and management.

Our clients include governments, local authorities and private companies.

Our goal is client's satisfaction.

Head Office
1-8-6 Nishiki, Naka-ku, Nagoya, Aichi Pref., Japan 460-0003
Tel:(81)-052-232-6053
Fax:(81)-052-221-7832
www.nakanihon.co.jp

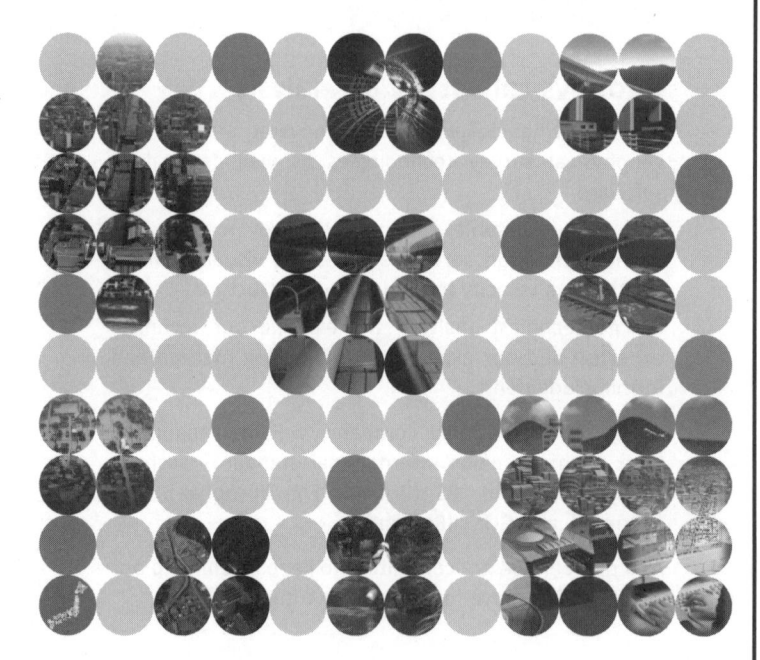

NAKANIHON
Engineering Consultants co.,ltd

NEWJEC
Enginnering Consultant

Linking Nature And People Through Technology

Fields of Activity
- Dam and Water Resources Management
- River, Erosion and Coastal Engineering
- Roads, Highways and Bridges, Transportation and New Traffic Systems
- Urban and Regional Planning
- Ports, Harbors and Airports
- Architectural and Structural Engineering
- Geology, Geotechnics and Foundations
- Electric Power Development Engineering
- Substations
- Information and Telecommunication Systems
- Development Planning, Environmental and Socioeconomic Engineering
- Land Formation and Landscaping, Water Supply and Sewerage Systems, Water Treatment and Disposal

http://www.newjec.co.jp

■**Osaka Head Office**
2-3-20 Honjo-Higashi, Kita-ku, Osaka 531-0074, Japan
Tel. +81-6-6374-4059   Fax. +81-6-6374-5198

■**Tokyo Head Office**
1-5-7 Kameido, Koto-ku, Tokyo 136-0071, Japan
Tel. +81-3-5628-7201   Fax. +81-3-5628-7200

■**Jakarta Office**
Summitmas I, 6th Floor, Jl. Jenderal Sudirman Kav. 61-62,
Jakarta Selatan, Indonesia
Tel. +62-21-2523460   Fax. +62-21-2523470

■**Myanmar Liaison Office**
Room No.427 c/o COCORO 4th fl. YUZANA Hotel, 130 Shwe Gon Taing
Rd. Bahan twp. Yangon, Myanmar
Tel.+95-1-860-4919

NIPPON KOEI — INTERNATIONAL ENGINEERING CONSULTANTS

- Japan's No.1 International Engineering Consultants
- Completed over 5,500 multi-disciplinary infrastructure projects in 160 countries during 70 years
- Oldest independent consulting firm in Japan, founded in 1946
- About **4,500** staff in Group
- Total paid-in capital US$ 70 million
- Listed on First Section of Tokyo Stock Exchange

Nippon Koei Worldwide Network of Offices

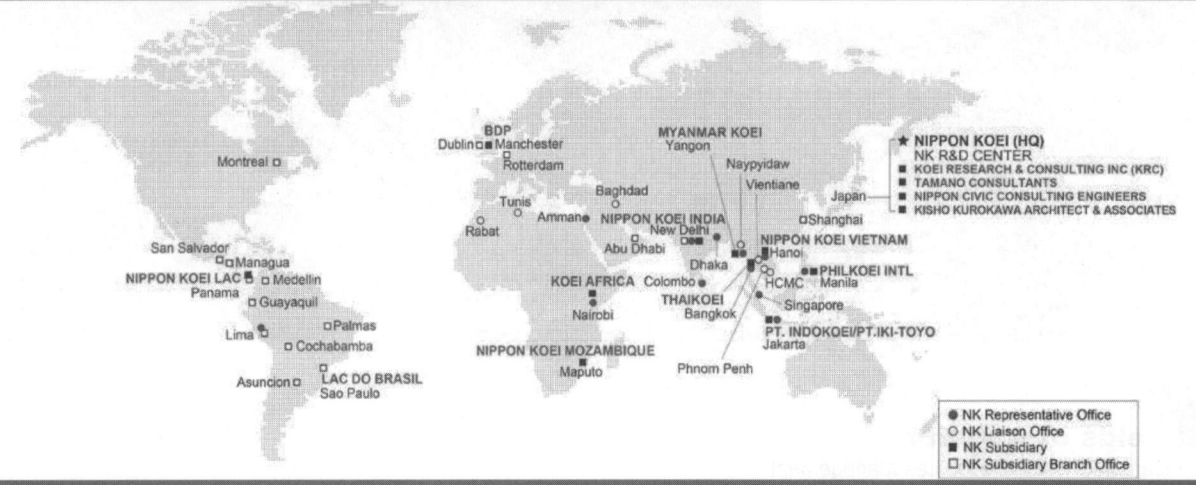

★ NIPPON KOEI (HQ)
- NK R&D CENTER
- KOEI RESEARCH & CONSULTING INC (KRC)
- TAMANO CONSULTANTS
- NIPPON CIVIC CONSULTING ENGINEERS
- KISHO KUROKAWA ARCHITECT & ASSOCIATES

Offices: Montreal, San Salvador, Managua, NIPPON KOEI LAC, Medellin, Panama, Guayaquil, Lima, Palmas, Cochabamba, Asuncion, LAC DO BRASIL Sao Paulo, BDP Dublin, Manchester, Rotterdam, Tunis, Rabat, Baghdad, Ammon, Abu Dhabi, KOEI AFRICA Nairobi, NIPPON KOEI MOZAMBIQUE Maputo, MYANMAR KOEI Yangon, Naypyidaw, Vientiane, NIPPON KOEI INDIA New Delhi, Dhaka, Colombo, Shanghai, NIPPON KOEI VIETNAM Hanoi, HCMC, THAIKOEI Bangkok, PHILKOEI INTL Manila, Singapore, PT. INDOKOEI/PT.IKI-TOYO Jakarta, Phnom Penh

- ● NK Representative Office
- ○ NK Liaison Office
- ■ NK Subsidiary
- □ NK Subsidiary Branch Office

Nippon Koei Research and Development Center

Environmental Science Lab.

Hydraulic Model Test Lab.

Automatic Tephra and Rain Gauge

Centrifuge

River Experiment Yard

Soil Laboratories Tests and Centrifuge of Nippon Koei R & D Center

We offer high quality technology services in the field of Geotechnical Engineering through our fully equipped facilities and technology skills to meet all your needs related to soil material, stability of the soil structure and special geotechnical problem such as rock-fill dam, highway, high fill at mountain airport, soft ground construction, disposal facilities and solid waste, soil-structure interaction problem, seismic safety, soil erosion geo-environmental problem etc..

Fully Equipped Facilities for Soil Material Tests

Permeability Test

Triaxial Compression Test(Standard, Middle, Large)

Compaction Test (Standard, Large)

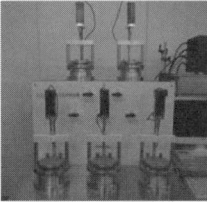

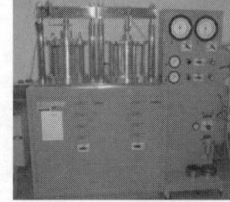

Consolidation Test(Standard, Middle, Large)

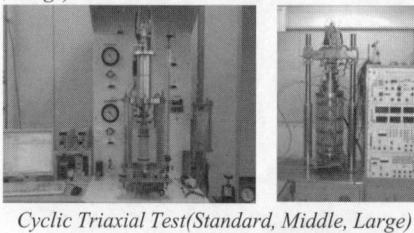

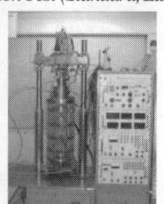

Cyclic Triaxial Test(Standard, Middle, Large)

Ring Shear Test

CBR Test

Standard Direct Box Shear Test (Static and cyclic shear)

Large Scale Direct & Simple Shear Test (Box: 40cm × 40cm)

Centrifuge Model Test

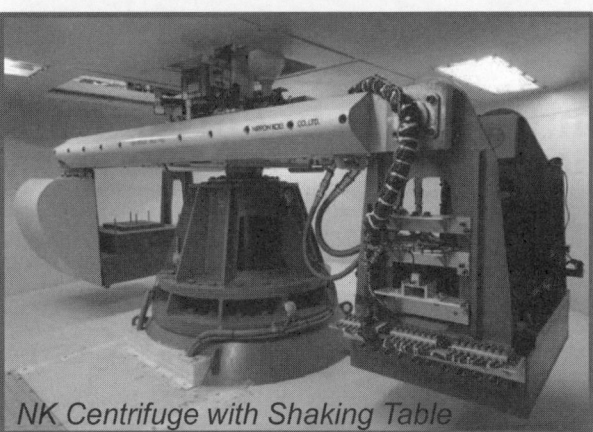

NK Centrifuge with Shaking Table

Specifications of Centrifuge and Related Facilities
Type : Beam   Effective Radius: R=2.6m
Max. Acceleration : 100 g   Max. Payload: 1,000 kg
Payload Capacity : 100 g-ton   Data Acquisition: 40 channels
Others: CCD camera, Digital Camera with Image analysis system

Specifications of Shaking Table
Shaking Control System: Electrohydraulic Servo Control
Max. Centrifugal Acceleration: 100 g
Max. Shaking Acceleration: 25 G (1/30 model 818gal)
Max. Payload: 250kg
Max. Displacement: ±3.0mm
Frequency Range: 10 - 400Hz

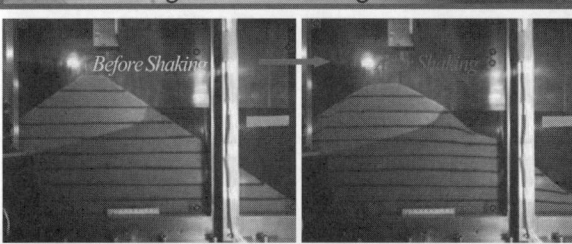

Before Shaking   *Shaking*   *Rigid soil container*   *Soil laminar shear box*

NIPPON KOEI CO., LTD.
(R&D Center)

2304, Inarihara, Tsukuba-city, Ibaraki 300-1259, JAPAN
TEL.+81-29-871-2000   FAX.+81-29-871-2019
E-mail:tkb-info@cx.n-koei.co.jp https://www.n-koei.co.jp/english

NITTOC Expertise

NITTOC can meet various kinds of social needs with special technologies cultivated through experience.

NITTOC offers various kinds of construction technologies that are taken account of the balance with the natural environments for urban redevelopment and the creation of new urban environments.

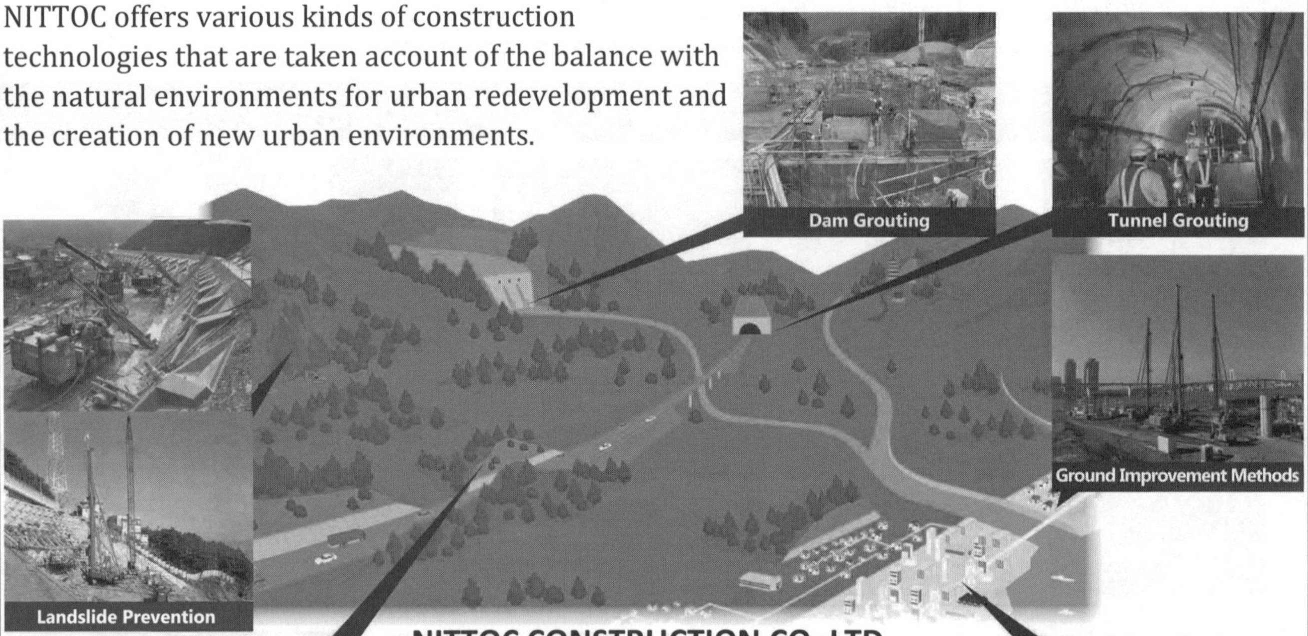

NITTOC CONSTRUCTION CO.,LTD.

HEADQUARTERS
4F, 5F and 6F, Daiwa Higashi Nihonbashi Bldg, 3-10-6,
Higashi-Nihonbashi, Chuo-ku, Tokyo 103-0004, Japan
TEL +81-3-5645-5050 FAX +81-3-5645-5051

JAKARTA Representative Office
GENERALI TOWER GRAND RUBINA BUSINESS PARK 16th
Floor Unit G, Kawasan Rasuna Epicentrum Jl. HR Rasuna Said,
Jakarta 12940, Indonesia
TEL.+62(21)2994-1582   FAX.+62(21)2994-1991

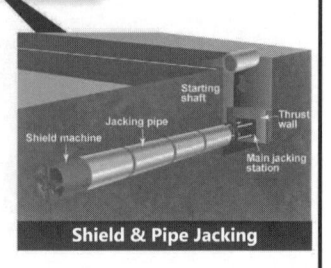

Seismic reinforcement by ground anchors in Port of Kobe

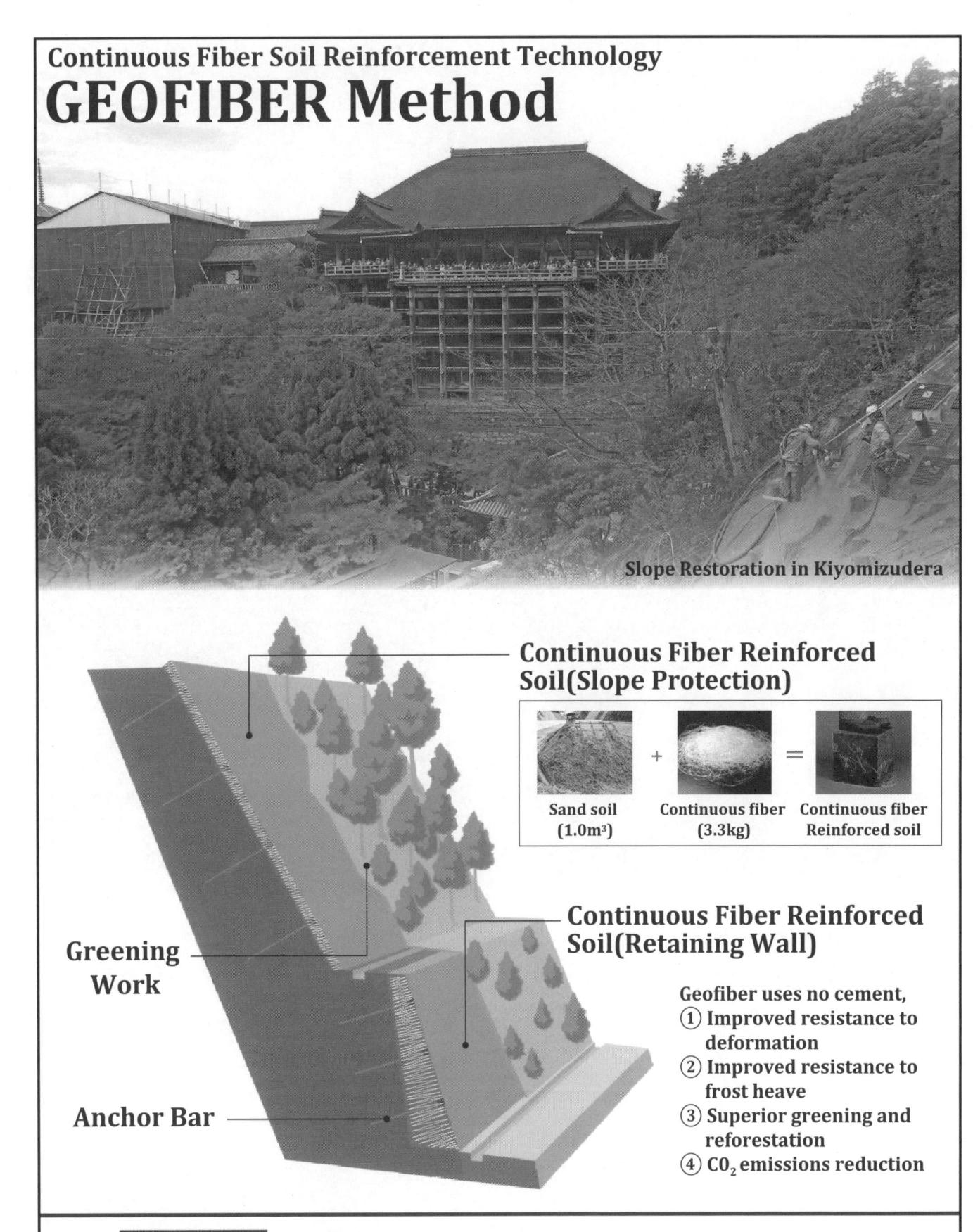

Continuous Fiber Soil Reinforcement Technology
GEOFIBER Method

Slope Restoration in Kiyomizudera

Continuous Fiber Reinforced Soil (Slope Protection)

Sand soil (1.0m³) + Continuous fiber (3.3kg) = Continuous fiber Reinforced soil

Continuous Fiber Reinforced Soil (Retaining Wall)

Greening Work

Anchor Bar

Geofiber uses no cement,
① Improved resistance to deformation
② Improved resistance to frost heave
③ Superior greening and reforestation
④ CO_2 emissions reduction

NITTOC NITTOC CONSTRUCTION CO., LTD.

HEADQUARTERS
4F, 5F and 6F, Daiwa Higashi Nihonbashi Bldg, 3-10-6, Higashi-Nihonbashi, Chuo-ku, Tokyo 103-0004, Japan
TEL +81-3-5645-5050 FAX +81-3-5645-5051

JAKARTA Representative Office
GENERALI TOWER GRAND RUBINA BUSINESS PARK 16th Floor Unit G, Kawasan Rasuna Epicentrum Jl. HR Rasuna Said, Jakarta 12940, Indonesia
TEL.+62(21)2994-1582   FAX.+62(21)2994-1991

Dubai Metro Project

Location: Middle East
Dubai, U.A.E
Facility type:
Railways/Airports/Harbor
Completion: August, 2011

Hoover Dam Bypass Project
- Colorado River Bridge

Location: North America
Nevada NV- Arizona AZ, U.S.A.
Facility type:
Roads/Tunnels/Bridges
Completion: October, 2010

Shaping the Times with Care

OBAYASHI CORPORATION

Shinagawa Intercity Tower B, 2-15-2, Konan,
Minato-ku, Tokyo 108-8502, Japan

http://www.obayashi.co.jp/english

OBAYASHI

A 100,000 kilometer tower connecting the earth to outer space

The Space Elevator Construction Concept

http://www.obayashi.co.jp/english/special/2014110424.html

Shaping the Times with Care

OBAYASHI CORPORATION

Shinagawa Intercity Tower B, 2-15-2, Konan, Minato-ku, Tokyo 108-8502, Japan

http://www.obayashi.co.jp/english

奥村組 Head Office: 2-2-2 Matsuzaki-cho, Abeno-ku, Osaka　TEL: +81-6-6621-1101
Tokyo Head Office: 5-6-1 Shiba, Minato-ku, Tokyo

Harmonize Human and Nature by Technology.

Respect human and nature, contribute to make the future better.
Such wishes being symbolized into the corporate logo of "人 (human)"
Meet people, create a town and an environment in harmony with people.
OKUMURA will step forward to be useful for human and nature through its construction business.

奥村組
OKUMURA CORPORATION
Head Office: 2-2-2 Matsuzaki-cho, Abeno-ku, Osaka　TEL: +81-6-6621-1101
Tokyo Head Office: 5-6-1 Shiba, Minato-ku, Tokyo　TEL: +81-3-3454-8111

http://www.okumuragumi.co.jp

OKUMURA's Pictorial Track Record

OKUMURA plays an active role in various fields toward a HUMAN CONSTRUCTER to make the future better by respecting human and nature.

build · extend · lay · construct

奥村組 OKUMURA CORPORATION
Head Office: 2-2-2 Matsuzaki-cho, Abeno-ku, Osaka TEL: +81-6-6621-1101
Tokyo Head Office: 5-6-1 Shiba, Minato-ku, Tokyo TEL: +81-3-3454-8111

http://www.okumuragumi.co.jp

OYO
oyo corporation

OYO is the leading geological / geotechnical investigation / consultation company in Japan. Since its establishment in 1957, we have been involved, with our quality services and instruments, in almost all major construction projects highlighted in Japanese economic growth, for buildings, dams, roads, tunnels, nuclear power plants…

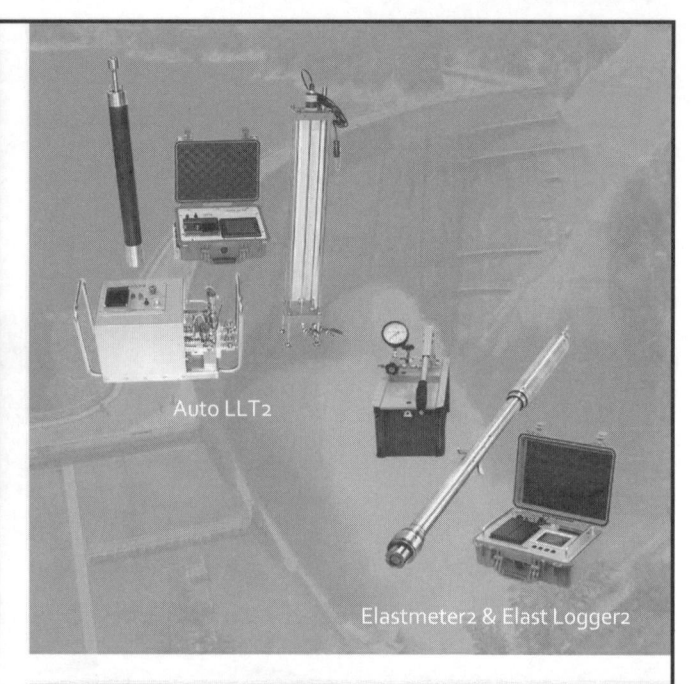

Auto LLT2

Elastmeter2 & Elast Logger2

Instruments & Solution

OYO is not only a service provider but also a manufacturer of advanced geotechnical and geophysical instruments. Our geo-expertise, both services and instruments, are currently being leveraged for the following business segments:

- Investigation & Inspection for Structures;
- Natural Disasters Management;
- Natural Environmental Preservation;
- Natural Resources Development.

Global Network of OYO Group

- Japan
- Malaysia
- China
- Guam
- Turkey
- France
- UK
- USA

http://www.oyo.com
international@oyonet.oyo.co.jp

i-SENSOR

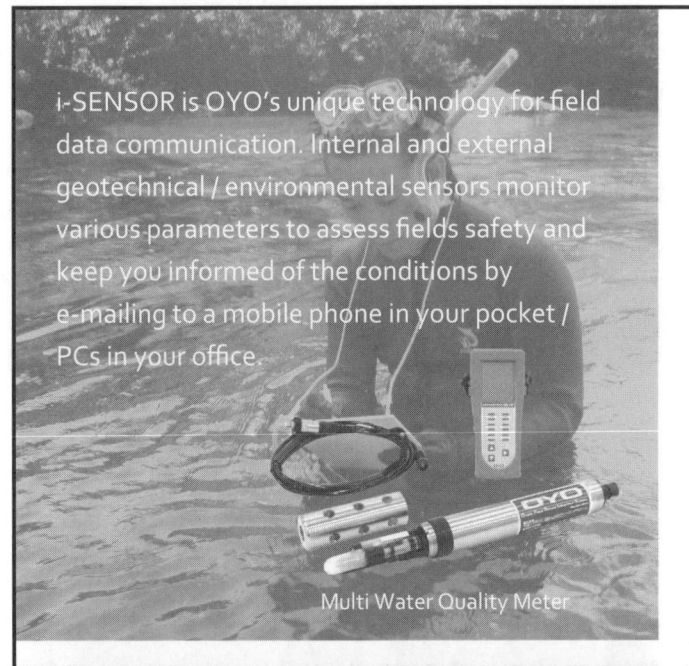

i-SENSOR is OYO's unique technology for field data communication. Internal and external geotechnical / environmental sensors monitor various parameters to assess fields safety and keep you informed of the conditions by e-mailing to a mobile phone in your pocket / PCs in your office.

Multi Water Quality Meter

How do you secure safe operation of your projects which are always exposed to risks of geohazards and environmental pollution? Land subsidence in tunnel construction, environmental pollution in dams, rock fall in roadside slopes... There are so many risks you have to monitor. How do you watch them when you cannot keep your eyes on them?

i-SENSOR will keep its eyes instead.

i-SENSOR is most tipically used to configure Early Warning System against geohazards. Combined with various range of our sensor products, the systems detect the occurrences timely and precisely.

Water Level Gauge S&DL mini

Stationary Borehole Inclinometer

i-SENSOR2
Remote & Real-Time Monitoring of Your Risk

i-SENSOR2 Tiltmeter

i-SENSOR2 Universal Logger

Nhật Tân Bridge
Vietnam-Japan Friendship Bridge

Project Outline

Employer:	PMU85, Ministry of Transport, Vietnam
Contractor:	JV of IHI Infrastructure Systems Co., Ltd. & Sumitomo Mitsui Construction Co., Ltd.
Total Length:	3,080 meter
Main Bridge Length:	1,500 meter
Width:	35.6 meter
Type of Structure:	(Main Bridge) 6-span continuous-steel girder-cable stayed bridge
	(Approach Bridge) Super T-girder / Prestressed concrete girder bridge
Construction Period:	October 2009 to December 2014

SUMITOMO MITSUI CONSTRUCTION CO., LTD.

Foundation Work
- Steel Pipe Sheet Pile -

*Piling Method developed in Japan
*Applied for the 1st time in Southeast Asia
*5 foundations built in Red River

SUMITOMO MITSUI CONSTRUCTION CO.,LTD.

Toward a Beautiful Age
The Tokyu Group

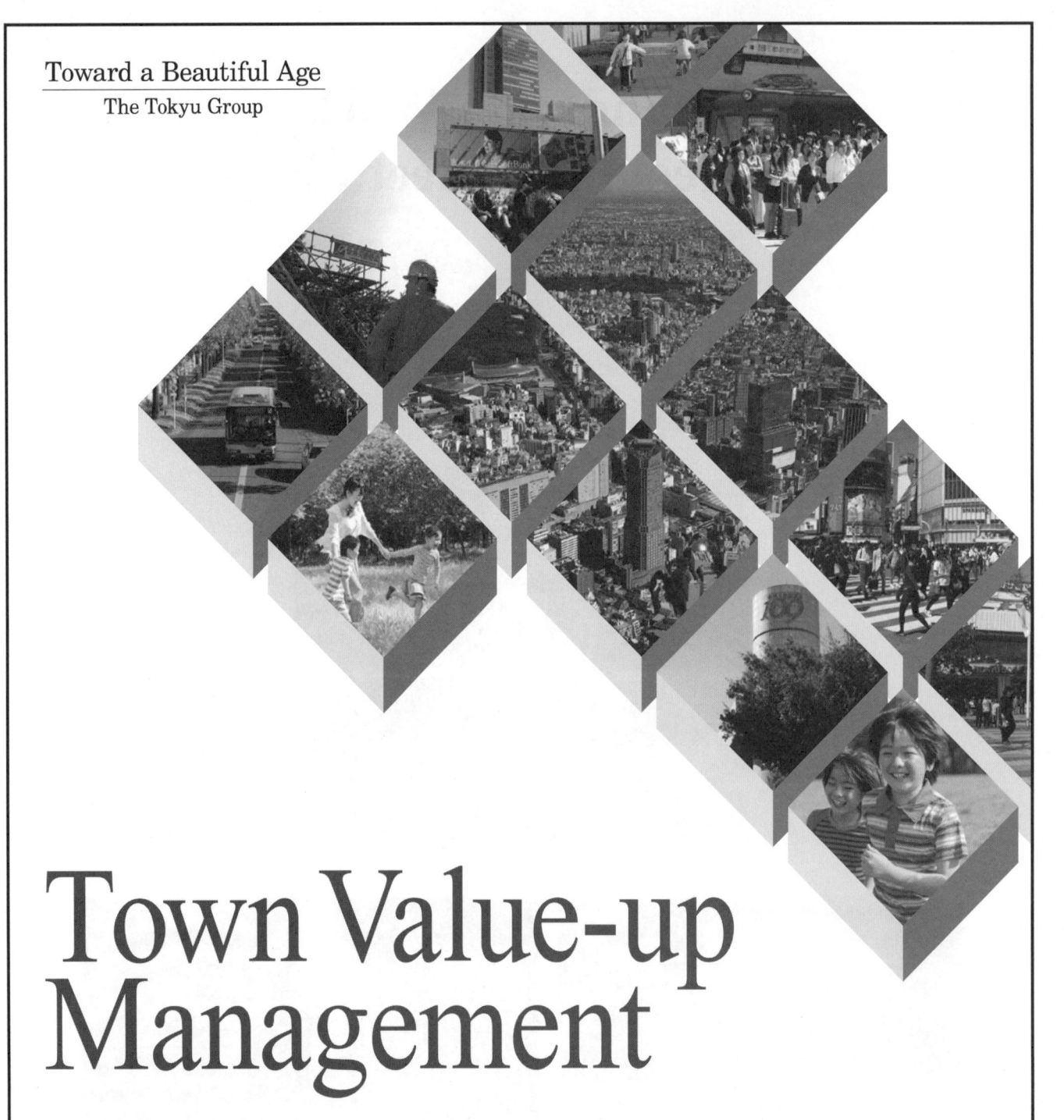

Town Value-up Management

Aiming at "New Value Creation" through construction,
we at Tokyu Construction consider the community
as a whole from the viewpoint
of customers and residents,
thereby contributing to the creation
of safe and pleasant living environments.

TOKYU CONSTRUCTION
http://www.tokyu-cnst.co.jp/english/

RRR-B method

-Reinforced Railroad with Rigid Facing Method-

RRR-B method is a construction method for railway and road embankment which is a composite Soil Structure of Geotextiles and RC walls. This method has an advantage compared to conventional embankment in the point that it can be applied in narrow spaces. Furthermore, it has been proven to have a high earthquake resistant performance and also applied to Japanese high speed rail way, "Shinkansen".

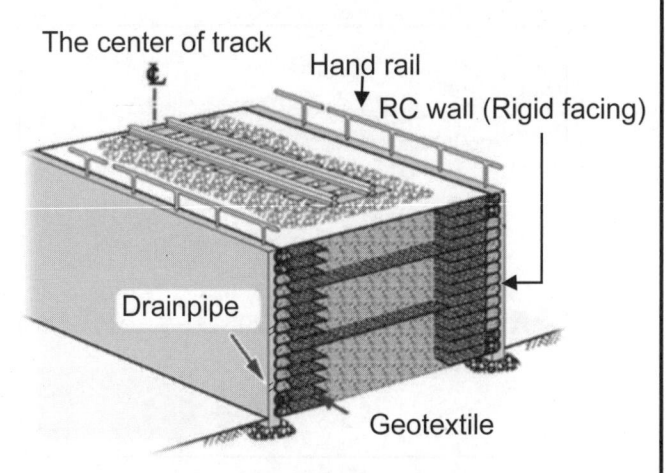

Overview of the RRR-B method

Radish anchor method

-Method of making large soil-cement anchor with high economical efficiency and workability-

The Radish anchor method is a method of constructing thick and short reinforcing materials inside an embankment. It can be applied in reinforcements against earthquakes and other disasters.

Diagram of the Radish anchor

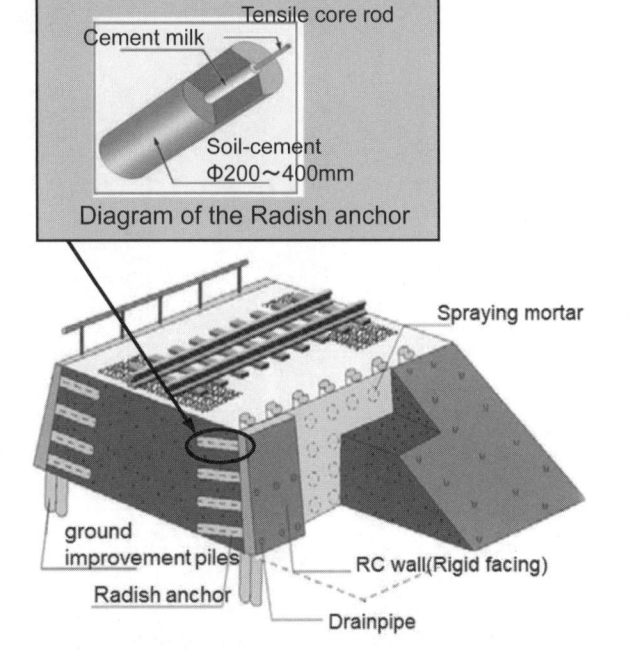

Overview of making a slope steeper

TOKYU CONSTRUCTION

http://www.tokyu-cnst.co.jp/english/technology/index.html

Giving Shape

to Dreams

Toyo Construction's strength lies in its unparalleled marine technologies. The company has grown to be one of Japan's leading marine contractors, with experience in the fields of dredging, reclamation and port construction and energy-related offshore structures.

Our start in operation for overseas business dates back to 1972 in Singapore, and has since established a solid reputation by completing a number of difficult marine and on land construction projects in Southeast Asia, the Middle East, and Pacific Ocean countries to expand to Kenya in Africa in 2012. Operation fields cover not only dredging and reclamation, ports, and energy-related facilities, but also bridges, dams, river, and building works.

With the rapid expansion of international economic blocs and the accelerated growth of globalization, overseas operations have become increasingly important. Making use of our cumulative expertise and technology, and working together with our large well trained overseas staff, Toyo Construction is committed to expand our operation fields even further.

TOYO CONSTRUCTION CO., LTD.

www.toyo-const.co.jp/en/

Rotary Vibration Drill
ECO-3V-3

Environmental Protection
High Speed, Low Noise,
Useful for Soil Investigation
- Rotation H:3.8 kN·m (Max)
  Torque L:1.9 kN·m (Max)
- Dimension 3,420 × 1,645 × 2,200 mm
- Gross Weight 2,650 kg

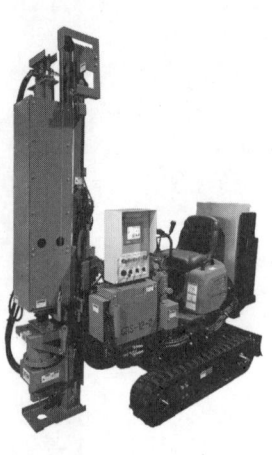

Automatic Ram Sounding Machine
CRS-12-2

The most accurate
ground strengthtesting for
deeper andharder formations.
- Pull up 39.6 kN
- Dimension 2,350 x 830 x 2,570 mm
- Gross Weight 1,630 kg

Dojyo Sukui
(Soil sampler)

Geological survey for mineral exploration, environment protection, and civil engineering

YBM offers worldwide reputed line up for geological surveying and drilling to meet with wide range of customers demands.

YBM Spindle Type Series

A complete line of drilling machines and the corresponding tools to meet broad needs. Such as small and large diameter, super light weight type and deep well drilling type.

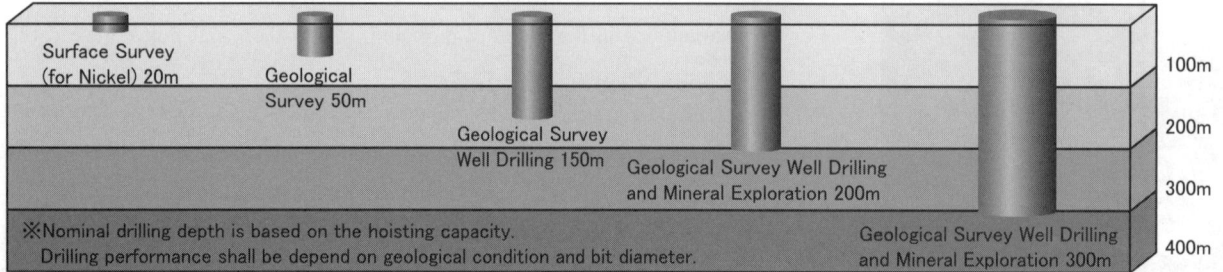

YHP-1	YBM-05DA-2	YBM-1WA	YBM-2WSⅡ	YBM-3JR
φ43 mm	φ43 mm(φ46 mm)	φ48 mm	φ93 mm(φ59 mm)	φ93 mm

- Surface Survey (for Nickel) 20m
- Geological Survey 50m
- Geological Survey Well Drilling 150m
- Geological Survey Well Drilling and Mineral Exploration 200m
- Geological Survey Well Drilling and Mineral Exploration 300m

※Nominal drilling depth is based on the hoisting capacity.
Drilling performance shall be depend on geological condition and bit diameter.

MODEL	YHP-1	YBM-05DA-2	YBM-1WA	YBM-2WSⅡ	YBM-3JR
Dimension (L×W×H)	1,030 x 740 x 1,525 mm	1,040 x 700 x 1,160 mm	1,550 x 700 x 1,310 mm	1,650 x 820 x 1,485 mm	1,560 x 860 x 1,700 mm
Approx. Weight	161 kg (Max.compo 26kg)	280 kg (Max.compo 27kg)	435 kg	810 kg	870 kg

YBM Co., Ltd. http://www.ybm.jp/

HEAD OFFICE:
1534, Haru, Karatsu, Saga Pref., 847-0031, Japan
Tel: (+81)955-77-1121 Fax: (+81)955-70-6010

TOKYO OFFICE:
2nd Fl., Yaesu Dai-San Nagaoka Bldg., 3-22-11 Hatchobori,
Chuo-ku, Tokyo, 104-0032, Japan
Tel: (+81)3-6280-4789 Fax: (+81)3-6280-4744

INDONESIA REPRESENTATIVE OFFICE:
Gedung Pusat Perfilman H. Usmar Ismail Lantai. 3
Ruang 343 JL. HR. Rasuna Said Kav.
C-22 Jakarta Selatan12940 Indonesia
Tel: (+62)21-5292-1131 Fax: (+62)21-5292-1132

Please feel free to contact us e-mail: welcome@ybm.jp

Ambitious for Technology & Knowledge

Groundwater Development

- Field Survey & Testing
- Aquifer Evaluation
- Well Design & Installation
- Well Operation & Maintenance

Geotechnical Engineering

- Field Boring & Testing
- Laboratory Testing
- Chemical, Numerical
- Structural & Remedial Action Design
- Monitoring System Development

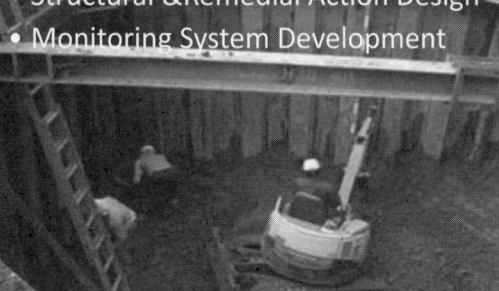

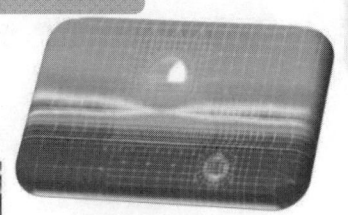

Structural Engineering

- Field Survey & Testing
- Seismic Capacity Analysis
- Reinforce Design
- Site supervision

ATK ASANO TAISEI KISO ENGINEERING CO., LTD.

Address/ TEL, E-mail	8-7 Kitaueno2chome, Taitou-ku, Tokyo, Japan, 110-0014 +81-3-5246-41721 atk-info@atk-eng.jp
Group Companies	ACKG Ltd. "Parent Company" (JASDAQ Listed) Oriental Consultants Co.,Ltd., Oriental Consultants Global Consultants Co.,Ltd A-Tec Co.,Ltd., Research & Solution

CHEMICAL GROUTING CO.,LTD.

- Seismic Reinforcement
- Liquefaction Mitigation
- Disaster Prevention
- Infrastructure Renewal
- Disaster Recovery
- Environmental Protection
- Social Infrastructure Development

CHEMICAL GROUTING CO.,LTD.

URL : http://www.chemicalgrout.co.jp

- **HEADQUARTERS**
  Kyodo News Bldg. 3F,2-2-5, Toranomon, Minato-ku, Tokyo, 105-0001 — Phone +81-3-5575-0511
- **Nishinihon Regional Office**
  Maruito OBP Bldg. 9F, 2-2-22, Shiromi, Chuo-ku, Osaka-shi, Osaka, 540-0001 — Phone +081-6-6946-7481
- **Taiwan Branch**
  No.9, Lane 65, Sec.2, 2F, Zhong-Shan N.RD., 104 Zhong-Shan Dist., Taipei Taiwan, R.O.C. — Phone +886-2-2522-9070
- **CGC Geotecnia e Construções LTDA**
  Rue Comendor Elias Assi, 63, Caxinqui, CEP:05516-000-São Paulo-SP-Brasil — Phone +55-11-2614-3363

Enriching life through engineering

Our mission is "enriching life through engineering".

As the first consulting engineering company in Japan,
we have been working with this philosophy for over 70 years.
Through our advanced technology and professional expertise,
we are committed to creating a prosperous society where people can
live with peace of mind.

CTI Engineering Co., Ltd.

3-21-1 Nihombashi Hama-cho Chuo-ku, Tokyo 103-8430 Japan
TEL:+81-3-3668-0451   E-mail:koho@ctie.co.jp
http://www.ctie.co.jp/english/

Engineering & Numerical Analysis Department Integrated Geotechnology Institute Limited (IGI)

We use geotechnologies to provide quality lifestyle infrastructure that is resilient against disasters.

IGI 株式会社複合技術研究所

Tel: 03-5368-4101, Fax: 03-5368-4105
Shinjuku district Yotsuya 1-23-6 TOKYO 160-0004
Tel: (+81)-3-5368-4101, Fax: +81-3-5368-4105
URL: http://www.igi.co.jp

Earth Doctor®

KGE serves as an "Earth Doctor" to benefit global society.

We provide the best solution through our own advanced technologies for the future.

- **Environment and Water Quality Survey**
- **Marine and Geophysical survey**
- **Disaster Prevention and Mitigation Management**
- **Infrastructure Maintenance Management**

Sincerely, Speedy and best Solution.
Kawasaki Geological Engineering Co., Ltd. http://www.kge.co.jp/

TEL: +81-3-5445-2071 FAX: +81-3-5445-2073 2-11-15, Mita, Minato-ku, Tokyo 108-8337 Japan E-mail: post-master@kge.co.jp

We build with heart.

People are concerned about the Earth.

Everyone wants to protect nature.

We are alive thanks to this planet's abundance.

This awareness is the basis for all of our actions.

Dedication to technology for the Earth.

A perpetual theme at Kumagai Gumi.

KUMAGAI GUMI
───── Building the future

A general contractor that our future will trust.

trust
of
the future

Maeda Corporation
http://www.maeda.co.jp

Successfully Building a Better Future.
NISHIMATSU CONSTRUCTION CO., LTD.

During the 50 years operating experience in Southeast Asia, we have worked closely together with the local business community and our local staff. In the process we have develop and establish a strong network of reliable and supportive business associates in these countries.

We shall continuously help in the development of these regional countries with our technological strength and good project management ability.

- Kwun Tong Line Extension Contract No.1001, Yau Ma Tei to Whampoa Tunnels and Ho Man Tin Station
- Aldrich Bay Development Work, Phase 4 
- Kaohsiung Metropolitan Mass Rapid Transit System 
- The Bang Pakong Diversion Dam 
- Paramount Bed Vietnam Factory Project 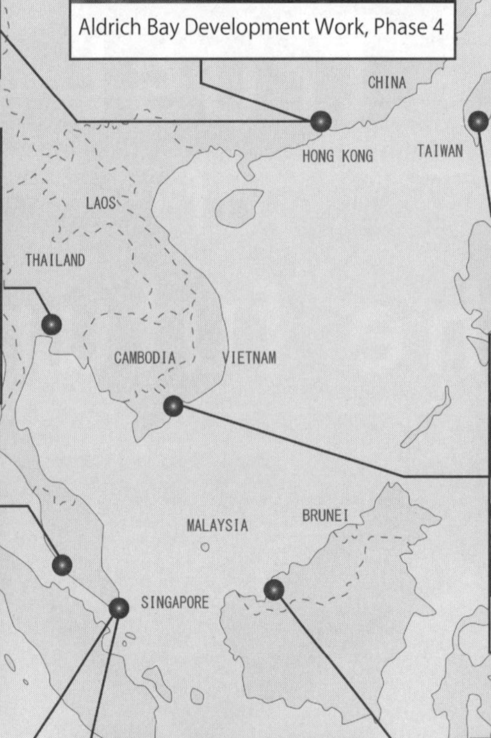 
- Pahang-Selangor Raw Water Transfer Project Lot 1-1 Water Transfer Tunnel and Related Works 
- Design, Construction & Completion of Stations at Nicoll Highway & Stadium Stations Including Tunnels
- Singapore National Library Construction 
- Centralized Sewerage System for Kuching City Centre (Package1)

西松建設
未来を創る現場力

GROUND IMPROVEMENT SPECIALISTS

ONODA CHEMICO
Onoda Chemical-Construction Since 1964

WE SUPPORT THE FUTURE

ONODA CHEMICO
Onoda Chemical-Construction Since 1964

- ONODA Jet Grouting Method
- ONODA Deep Mixing Method
- Lightweight Embankment Method
- Chemical Grouting Method

HEAD OFFICE
3-21 Kandanishiki-cho, Chiyoda-ku, Tokyo, 101-0054, Japan
TEL.+81-3-6386-7051 / FAX +81-3-6386-7021

VIETNAM OFFICE
Suite 3C, 34th Floor, Bitexco Financial Tower, 2 Hai Trieu, Ben Nghe Ward, District 1, Ho Chi Minh City, Vietnam
TEL.+84-8-3821-1319 / FAX +84-8-3821-1320

MYANMAR OFFICE
Unit 428, Yuzana Hotel, 4th Floor, No.130, Shwe Gon Taing Road, Bahan Township, Yangon, Myanmar
TEL.+95-1-8604920

- **Mechanical Mixing**
  Deep Soil Mixing
  (Wet Mixing, Dry Mixing)
  Mass Stabilization
  (WILL Method)

- **Chemical Grouting**
- **Jet Grouting**
  (V-JET Method)
- **Compaction Grouting**
  (CPG Method)

- **Ground Anchors**
- **Soil Nailing**
- **Micro Piling**

Our Mission: Providing the most economical and practical solutions.
An Innovator: Challenging sprit in geotechnical industry.
Superior Performance: Huge experiences for 60 years.

SANSHIN is a Japanese top company to have been working on reinforcement and stabilization of the ground since 1956. Our superior technology contributes in the world with building safe and comfortable society.

SANSHIN CORPORATION

URL : http//www.sanshin-corp.co.jp
E-mail : sales@sanshin-corp.co.jp

Headquarters	2-19-6 Yanagibashi, Taito-ku, Tokyo 111-0052, Japan	+81-3(5825)3700
Taipei Branch	5F-2, 130 Po Ai Road Taipei, Taiwan, R.O.C 10043	+866-2(2388)8039
Hong Kong Branch	Room 1106, the L.plaza 367-375, Queen's Road, Central, Hong Kong	+852(2666)0088

Link People, Connect Towns and Extend to the Future

信用と技術の
TEKKEN 鉄建

tekken global

About Tekken

Tekken Corporation was established in 1944 as Tetsudo Kensetsu Kogyo Kabushiki Kaisha and began a step as a company to do the infrastructure development chiefly composed of railway-related works. Since its establishment, Tekken has built relationships of trust with stakeholders including customers and is widely contributing to the development of a society based on technological strength.

Business of Tekken Corporation

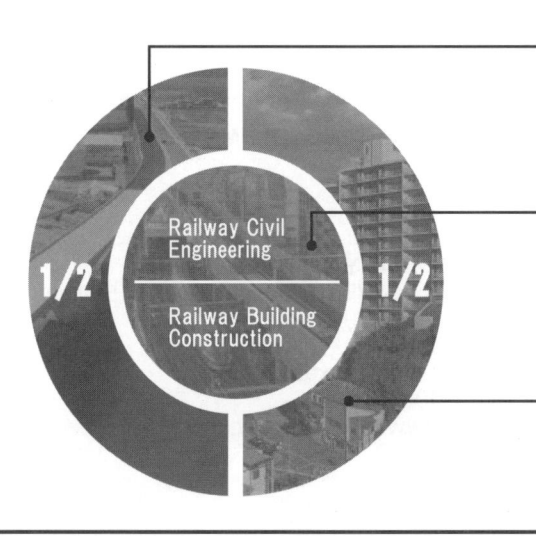

1/2 Railway Civil Engineering
1/2 Railway Building Construction

Civil Engineering

Business that develops infrastructure to support people's life, which includes road, tunnel, bridge, waterworks, sewerage and river revetment, etc.

Railway

Railway business has two fields, railway civil engineering to develop grade separation, tunnel, bridge and embankment for railway, and railway building construction to create station shed, station building and depot.

Building Construction

Business that creates buildings for the purpose that people live in comfort daily, which includes building, condominium, school and medical welfare facility, etc.

We are Pioneer in Soil Improvement

ⅩⅩ TENOX CORPORATION

TENOCOLUMN

TENOX took on the challenge of developing a high-quality soil improvement technique and we successfully established the soil improvement technique applicable to construction of foundations for buildings for the first time in Japan.

TENOCOLUMN can be used high concentration cement milk (W/C 45 to 50%) than other general methods (W/C 60 to 80%) by using special secondary additive.

As a result, 1) High strength 2) Low waste industry has been realized.

Figure 1 TENOCOLUMN (Max 2,600mm Dia.)

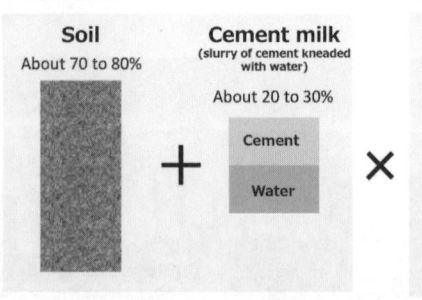

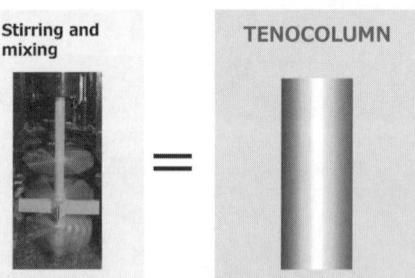

■ Material　　■ Construction　　■ Finished product

Soil About 70 to 80% + Cement milk (slurry of cement kneaded with water) About 20 to 30% (Cement / Water) × Stirring and mixing = TENOCOLUMN

Figure 2 Principle of TENOCOLUMN Construction Method

Figure 3 Base Machine

Application of TENOCOLUMN

TENOCOLUMN has been applied to about 35,000 projects in Japan (Buildings / Civil works).
For Buildings　: Detached houses, apartments, government buildings, skyscrapers, large logistics warehouses, airport, power plant etc. (Fig.4).
For Civil works : Foundation of tower and retaining wall, prevention of settlement, earth retaining, arc slip prevention etc. (Fig.5).

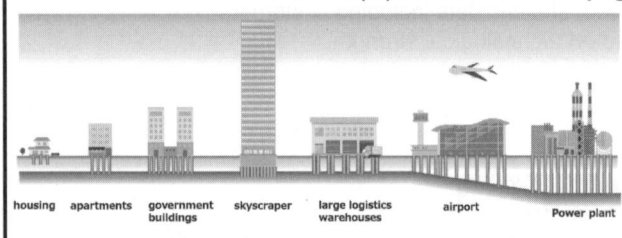

Figure 4 Application of TENOCOLUMN for Buildings

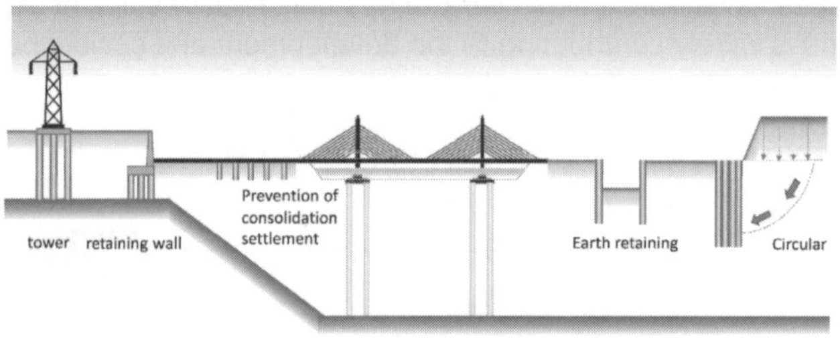

housing　apartments　government buildings　skyscraper　large logistics warehouses　airport　Power plant

Figure 5 Application of TENOCOLUMN for Civil works

tower　retaining wall　Prevention of consolidation settlement　Earth retaining　Circular

Contact Info.

TENOX CORPORATION
Office　　: Overseas Business Group New Business Promotion Dept., 4th Floor, Hulic Mita Building, 5-25-11, Shiba, Minato Ward, Tokyo, Japan
E-mail　 : ginfo@tenox.co.jp
Tel.　　　: +81-3-3455-7790
URL　　 : http://www.tenox.co.jp

TENOX ASIA COMPANY LIMITED
Office　　: 5th Floor, AVENIS Building, No.145 Dien Bien Phu St., Dakao Ward, Dist. 1, Ho Chi Minh City, Vietnam
E-mail　 : business@tenoxasia.com
Tel.　　　: +84-902-446-426
URL　　 : http://www.tenoxasia.com

To Future of Dear Earth

From Professional Group of Laboratory Tests for Geomaterials

KG&ERC Kansai Geo and Environment Research Center

Address : 1-3-3 Higashi-befu, Settsu, Osaka, Japan
Post code : 566-0042
Telephone : +81-6-6827-8833
E-mail : tech@ks-dositu.or.jp
URL : http://ks-dositu.or.jp/

Osaka Japan

Geo-Material Testing Center for Soil, Rock & Concrete
Hokkaido Soil Research Co-operation

Researching Geo-Materials & Pioneering a New Future
Corresponding to the Projects of Construction, Disaster Prevention & Environmental Conservation

Particle Size Distribution Test of Soil	Consolidation Test of Soil	Triaxial Compression Test of Soil
X-ray Diffraction Method	Cyclic Triaxial Test of Soil (Liquefaction Strength · Deformation Characteristics)	
Compression Test of Concrete	Laboratory Assessment for Deterioration of Concrete (Carbonation · Alkali Aggregate)	

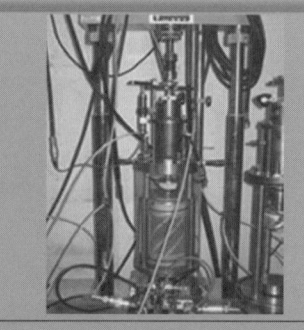

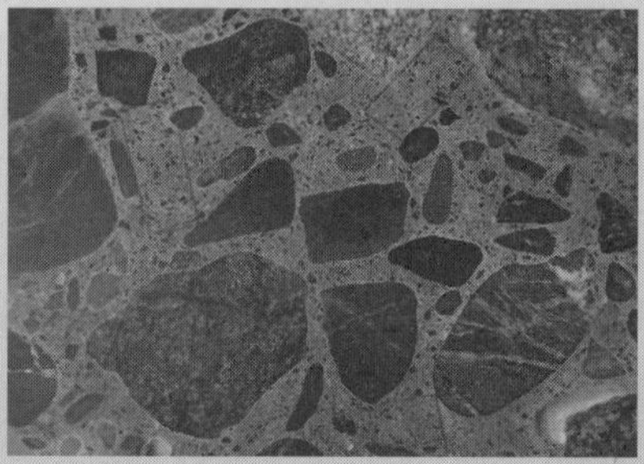

SRC

ISO/IEC 17025 Accredited Testing Laboratory
Hokkaido Soil Research Co-operation

Sapporo Japan

Address : 1-8-3-1, Kitago, Shiroishi-ku, Sapporo, Hokkaido Japan
Post code : 003-0831
Telephone : +81-11-873-9895
E-mail : info@src.or.jp
URL : http://www.src.or.jp

Chuden Engineering Consultants
A multi-disciplinary consulting services provider

Our service areas

- River / Debris-slide protection
- Civil engineering machine design
- Road / Transportation
- Architecture
- Harbor
- Environment
- Soil remediation / waste treatment
- Geology
- Regional arrangement
- Survey
- Water and sewer services
- Facility design / investigation
- Electric / communication
- Numerical analysis
- Information system

Energia
The chugoku electric Power group

Contact:
TEL: +81-82-255-5501
E-mai : inquir@cecnet.co.jp
URL : http://www.cecnet.co.jp

We meet the needs of our customers with creativity and comprehensive knowledge

We sincerely intend to meet social needs and carry out our mission to deliver infrastructure to the next generation beyond the current level under the corporate motto as "Sincerity, Brightness & Solidarity".

http://www.cfk.co.jp/en/

CHUO FUKKEN CONSULTANTS CO., LTD.
CREATIVE & FULL KNOWLEDGE

[Head Office]
4-11-10,Higashinakajima Higashi Yodogawa-ku,Osaka,533-0033,JAPAN
TEL:+81-6-6160-1227   FAX:+81-6-6160-1239

[Tokyo Head Office]
2-10-13,Kojimachi,Chiyoda-ku,Tokyo,102-0083,JAPAN
TEL:+81-3-3511-2002   FAX:+81-3-3511-2032

Upgrading and disseminating coastal and ocean technology
- toward the improvement of people's life and to ensure the safety and security -

Coastal Development Institute of Technology

SHIMBASHI SY BLDG., 5F, 1-14-2, NISHISHIMBASHI,
MINATO-KU, TOKYO 105-0003, JAPAN
TEL: 81-3-6257-3701, FAX: 81-3-6257-3706
E-mail: cdit0927@cdit.or.jp, URL: http://www.cdit.or.jp/

Infrastructure Solution Consultant

Eight-Japan Engineering Consultants Inc.

Valuable Future Environment

Contribution to the Creation of a Truly Affluent Society through Excellent Earth-Friendly Technologies and Professional Judgement

■ **Corporate Headquarters**
1-21 Tsushima Kyomachi 3-chome, Kita-ku, Okayama City,
Okayama, Japan 700-8617
TEL: +81 (0)86-252-8917 FAX: +81(0) 86-252-7509

■ **Tokyo Headquarters (International Department)**
33-11 Honcho 5-chome, Nakano-ku, Tokyo, Japan 164-8601
TEL: +81(0) 3-5341-5155 FAX: +81(0) 3-5385-8510

■ **Bangkok Representative Office**
B.B. Building, 15th Floor, Room 1515, 54 Sukhumvit 21 Road (Asoke),
Klong toey Nua, Wattana, Bangkok 10110 Thailand
TEL: +66 (0)2-664-4144, 4146 FAX: +66 (0)2-664-4145

URL: http://www.ejec.ej-hds.co.jp

FUJITA
Daiwa House Group®

http://www.fujita.co.jp 4-25-2 Sendagaya, Shibuya-ku, Tokyo 151-8570, Japan Phone: +81-3-3402-1911

変革の時代を超えて豊かな未来を創る…
「未来社会創造企業」

Survey
Design
Infrastructure management

復建調査設計株式会社
FUKKEN CO., LTD.

2-10-11, Hikari-machi, Higashi-ku, Hiroshima-shi, Hiroshima, 732-0052 JAPAN
TEL +81-82-506-1811   FAX +81-82-506-1890
URL http://www.fukken.co.jp/

■joint-venture company■

Fukken & Minami Consultant Co., Ltd.

Green Building, 540/1 Cach Mang Thang Tam Street, Ward 11. District 3, HCM City. VIETNAM
TEL +84 8 5404 5343   FAX +84 8 5404 5344
URL http://fmcvietnam.net/

With people and technology,

We challenge the future.

General Contractors, Architects & Engineers

HAZAMA ANDO CORPORATION

http://www.ad-hzm.co.jp

HOJUN
HOJUN CO.,LTD.

Address:1433-1 Haraichi,Annaka-shi,Gunma 379-0133   TEL:+81-27-385-3411 (front desk)   FAX:+81-27-385-5859

Hojun Group is a unique corporation that mines and mills bentonite—which is said to have thousands of applications—and develops it into products, on a scale unmatched by any other corporation in Japan. Hojun`s domestic share of the bentonite market is constantly above20%. We maintain an excellent international in reputation as one of the 10 largest manufacturers in the world.
At present, we have facilities capable of producing 11,000 tons of bentonite(domestic production only) per month. Production is expected to grow even more by importing high-quality foreign bentonite and mixing it with bentonite from the United States.

The NATURAL BLANKET METHOD is a construction method that layers 100% pure bentonite (crushed rock), a natural clay mineral.
The method provides high-performance seepage control (a water permeability coefficient of 1×10^{-10} m/s or lower)
Along with cesium absorption, forming a firm water Shut-off layer for various purposes such as the isolation of Radioactive contaminants and seepage control work in landfills.

exhibits an outstanding capacity for seepage control by means of mixing and compacting bentonite with soil.
This method is widely used for the containment of surplus soil with hazardous substances including heavy metals or seepage control technology for landfill bottom liners and capping.

HOKKOKU CHISUI Co., Ltd.

Boiling passion like the earth
Gushing out sensibility like water
Here are joys to be united with nature.

Main works
- Geotechnical survey
- Hydrological survey
- Soil tests
- Drilling of wells
- Landslide prevention
- Slope stability
- Inspection of civil engineering structures

Contact
25-1 Mikage, Kanazawa City,
Ishikawa, 921-8021, Japan
TEL: +81(0)76-241-7158
FAX: +81(0)76-243-2422
URL. http://www.hokukoku.co.jp

We strive for a better & safer society
through effective and innovative geotechnical solutions.

JCE Network
Geological Survey/Geotechnical Investigation/
Environmental Survey/Landslide Hazard Management/
Slope Hazard Management/Volcano/Seismic Hazard Management/
Avalanche Control/River Management/Road/Bridge/Tunnel Management/
Forest Management/Rural Development/Coastal Management/
Information Technology/ GIS/Rural Planning/ Permits/Simulation

ISO 9001 **JAPAN CONSERVATION ENGINEERS & CO., LTD.**

3-18-5 Toranomon Minato-Ward Tokyo,105-0001 JAPAN   TEL +81-3-34363673   FAX +81-3-34323787   URL:http://www.jce.co.jp/

Portable Bearing Tester "CASPFOL"

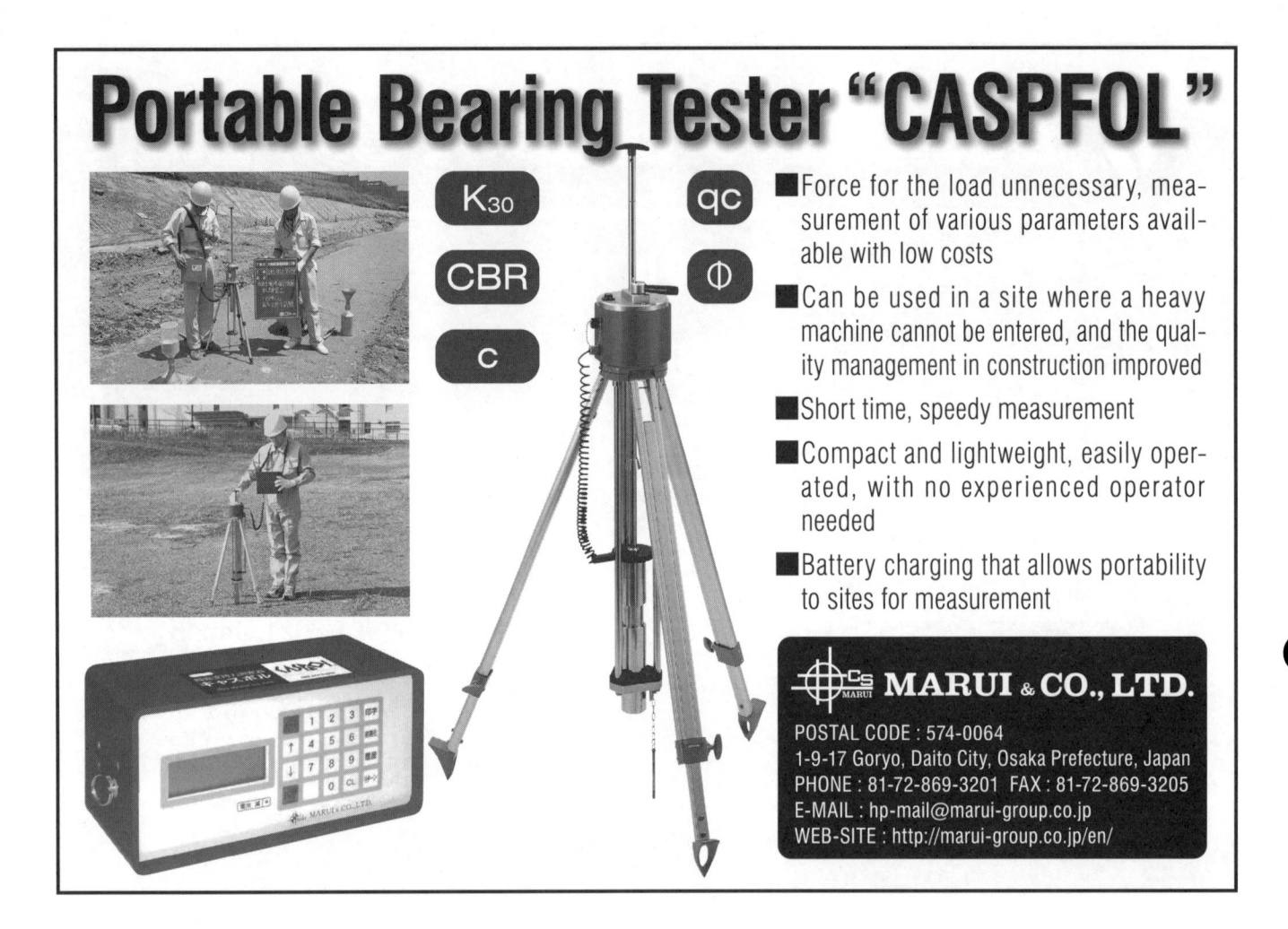

K_{30} q_c

CBR ϕ

c

- Force for the load unnecessary, measurement of various parameters available with low costs
- Can be used in a site where a heavy machine cannot be entered, and the quality management in construction improved
- Short time, speedy measurement
- Compact and lightweight, easily operated, with no experienced operator needed
- Battery charging that allows portability to sites for measurement

MARUI & CO., LTD.

POSTAL CODE : 574-0064
1-9-17 Goryo, Daito City, Osaka Prefecture, Japan
PHONE : 81-72-869-3201 FAX : 81-72-869-3205
E-MAIL : hp-mail@marui-group.co.jp
WEB-SITE : http://marui-group.co.jp/en/

---

nikken.jp

more than creative

Planners Architects Engineers

Geotechnical & Environmental Engineering
- Reclamation subsidence analysis
- Dredging disposal recycling & treatment
- Earthquake resistance design
- Water quality assessment
- Flood & Tsunami disaster area analysis

Industrial Facility Infrastructure Design
Urban Development & Infrastructure Design
Regional & Urban Development Planning

NIKKEN SEKKEI CIVIL ENGINEERING LTD
http://www.nikken-civil.co.jp/eng/

1-4-27 Koraku, Bunkyo-ku, Tokyo 1120004, JAPAN
Tel. +81-3-5226-3070 Fax. +81-3-5226-3075

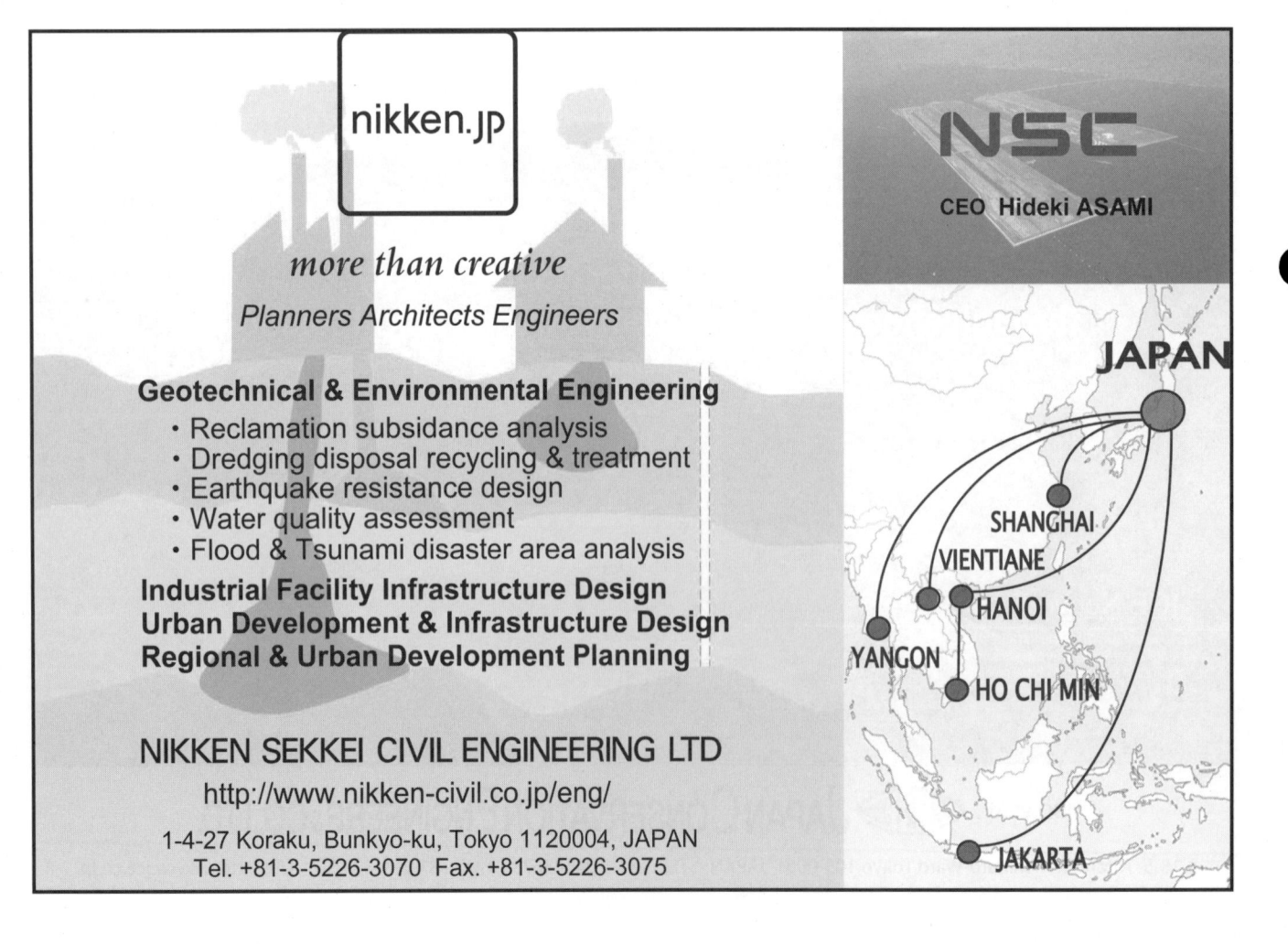

NSC
CEO Hideki ASAMI

JAPAN
SHANGHAI
VIENTIANE
HANOI
YANGON
HO CHI MIN
JAKARTA

CREATION of NiX

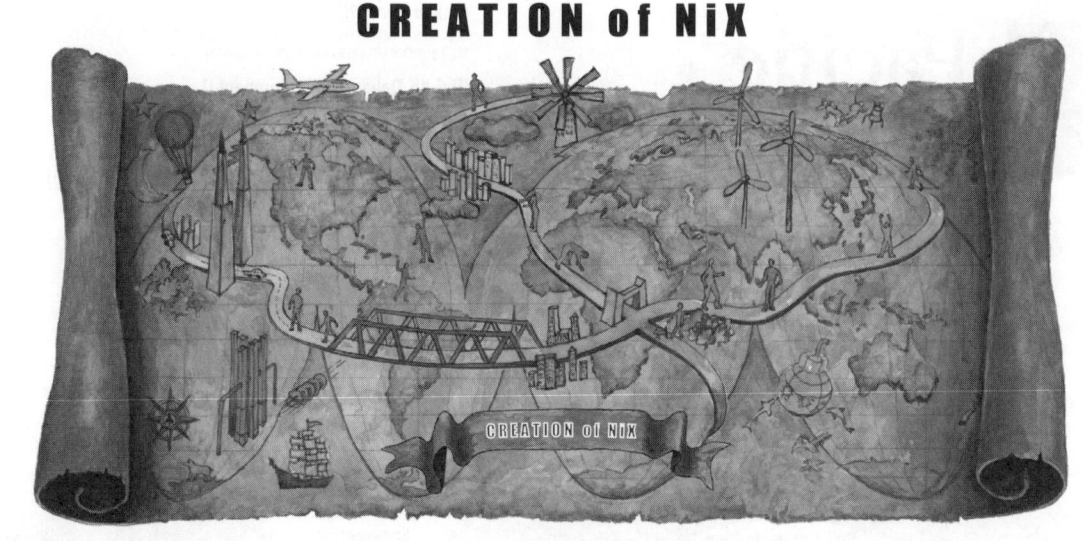

NiX [niks] Co., Ltd.

Construction consultant firm
- Design and planning of social infrastructure
- Survey · compensation · Information system

Head Office : 910-1 Yoshizukuri, Toyama-city, Toyama 930-0142, JAPAN   TEL: +81-76-436-2111   FAX: +81-76-436-3050
Tokyo Head Office : 6-1-1 Higashi-Ueno, Taito-ku, Tokyo 110-0015, JAPAN   TEL: +81-3-6802-8876   FAX: +81-3-6802-8626

Group companies
- NiX New Energy co.,ltd.
- NiX Holdings Singapore Pte.,Ltd
- Fields co.,ltd.
- PT. Lebong Sukses Energi

http://www.shinnihon-cst.co.jp

OCDI
Supporting the development of ports, people and technology

The Technical Standards and Commentaries for Port and Harbour Facilities in Japan

An English version of the Japanese Standards for port and harbor facilities can be downloaded at http://www.ocdi.or.jp/en/

Main Contents are;
- Principle of Performance-based Design Method
- Meteorology and Oceanography · Natural Conditions
- Material · Foundations · Waterways and Basins
- Protective Facilities for Harbors · Mooring Facilities
- Storage Facilities · Other Facilities

What is OCDI

The Overseas Coastal Area Development Institute of Japan (OCDI) was established in 1976 for the purpose of sharing Japan's experience and knowledge in port development with developing countries, and has since contributed to the economic development of countries throughout the world.
For more information, you are referred to our web-site http://www.ocdi.or.jp/en/

Pacific Consultants

[Tokyo Head Office]
3-22 Kanda-Nishikicho, Chiyoda-ku, Tokyo, Japan 1018462
TEL:+81(0)3-6777-1759 FAX:+81(0)3-3296-0522

[P.C.KK Dalian(Group Company)]
B-2F,32 Lixian Street, High-tech Industrial Zone, Dalian City, Liaoning Province, China 116023
TEL:+86(0)411-84755015 FAX:+86-(0)411-84753006

[PCKK International Asia Pte. Ltd.(Group Company)]
250 North Bridge Road #16-02, Raffles City Tower, Singapore 179101
TEL:+65-6265-1708 FAX:+65-6265-1758

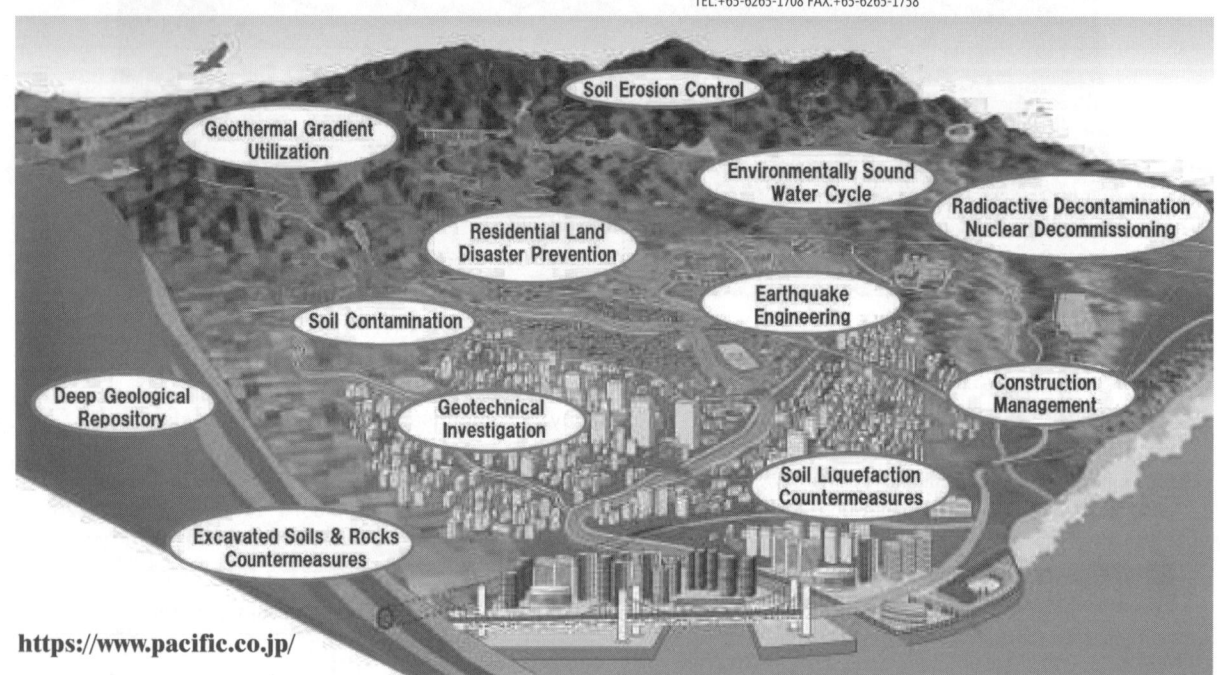

- Soil Erosion Control
- Geothermal Gradient Utilization
- Environmentally Sound Water Cycle
- Radioactive Decontamination Nuclear Decommissioning
- Residential Land Disaster Prevention
- Soil Contamination
- Earthquake Engineering
- Deep Geological Repository
- Geotechnical Investigation
- Construction Management
- Soil Liquefaction Countermeasures
- Excavated Soils & Rocks Countermeasures

https://www.pacific.co.jp/

Since its establishment in 1896 in Kure City, Hiroshima Prefecture, Penta-Ocean Construction Co., Ltd. has grown with society by contributing an enterprising, up-and-coming spirit and leading-edge construction technologies.
Today, a mentality of continually seeking challenges in new fields remains part of our corporate DNA.
A spirit of accepting challenges that never varies, even as times change, and the power of flexible self-innovation to respond to the needs of each new era.
At Penta-Ocean, we are never satisfied with things as they are, and we continue to move steadily forward, step by step.

GOING FURTHER
その先の向こうへ

2-2-8 KORAKU,BUNKYO-KU,TOKYO 112-8576 JAPAN
http://www.penta-ocean.co.jp/

PENTA-OCEAN CONSTRUCTION CO.,LTD.

Total Project Excellence

Expounding on Sato Kogyo's guiding corporate philosophy of 'Total Project Excellence', we have strived in years past to enhance customer satisfaction, build safe, secure and comfortable spaces and develop high-quality social infrastructure. Going forword, we remain dedicated to placing our customers' interests first, responding to society's needs, maintaining excellence and being a good partner to high-tech service organizations.

Once Built
never forgotten

General Contractor
SATO KOGYO CO.,LTD.
Founded 1862
www.satokogyo.co.jp

SCOPE

We provide a wide variety of information and know-how from planning, cost estimation, ordering, construction to maintenance and management for the smooth implementation of port and airport construction projects.

Name   SCOPE (Service Center of Port Engineering)
Corporation approved by Prime Minister.

Date of Establishment   May 30, 1994.
(Transformed into General Incorporated Foundation on April 1, 2013.)

Head Office   Shouyu Kaikan 3F, 3-3-1 Kasumigaseki, Chiyoda-ku, Tokyo 100-0013
TEL. +81-3-3503-2081 ／ FAX. +81-3-03-5512-7515

URL : http://www.scopenet.or.jp

Field density-moisture meter for compaction of soil

ANDES

1. You can judge a state of the compaction quickly by using ANDES.
2. ANDES can apply to various quality of soil.
3. Speedy measuring in only 1 min.
4. Safety and easy because using low level radiation source.
5. Having calculation function for data analysis.

Specification
Type　Densitty: Transmitted Gamma ray
　　　Moisture: Transmitted Fast neutron ray
Radiation source: Completely sealed capsule
　　^{60}Co2.59MBq & ^{252}Cf1.11MBq
Dimension: 340W×260D×145H, Weight 10.5kg

Others ・SRID・COARA・PIRICA.　etc.

ソイルアンドロックエンジニアリング株式会社
Soil and Rock Engineering Co.,Ltd.
URL : http:/www.soilandrock.co.jp
E-mail : sre@soilandrock.co.jp
Head office : 2-21-1 Shonaisakaemachi, Toyonaka, Osaka, Japan 561-0834
　　TEL (81)6-6331-6031　FAX (81)6-6331-6243

私たちは、
未来に誇れる環境を
これからも創造し続けます。

人と地球の架け橋に　竹中土木

本　社　〒136-8570　東京都江東区新砂1丁目1-1　電　話　03-6810-6200
WEB　https://www.takenaka-doboku.co.jp

Dreams into Reality for a Sustainable Future

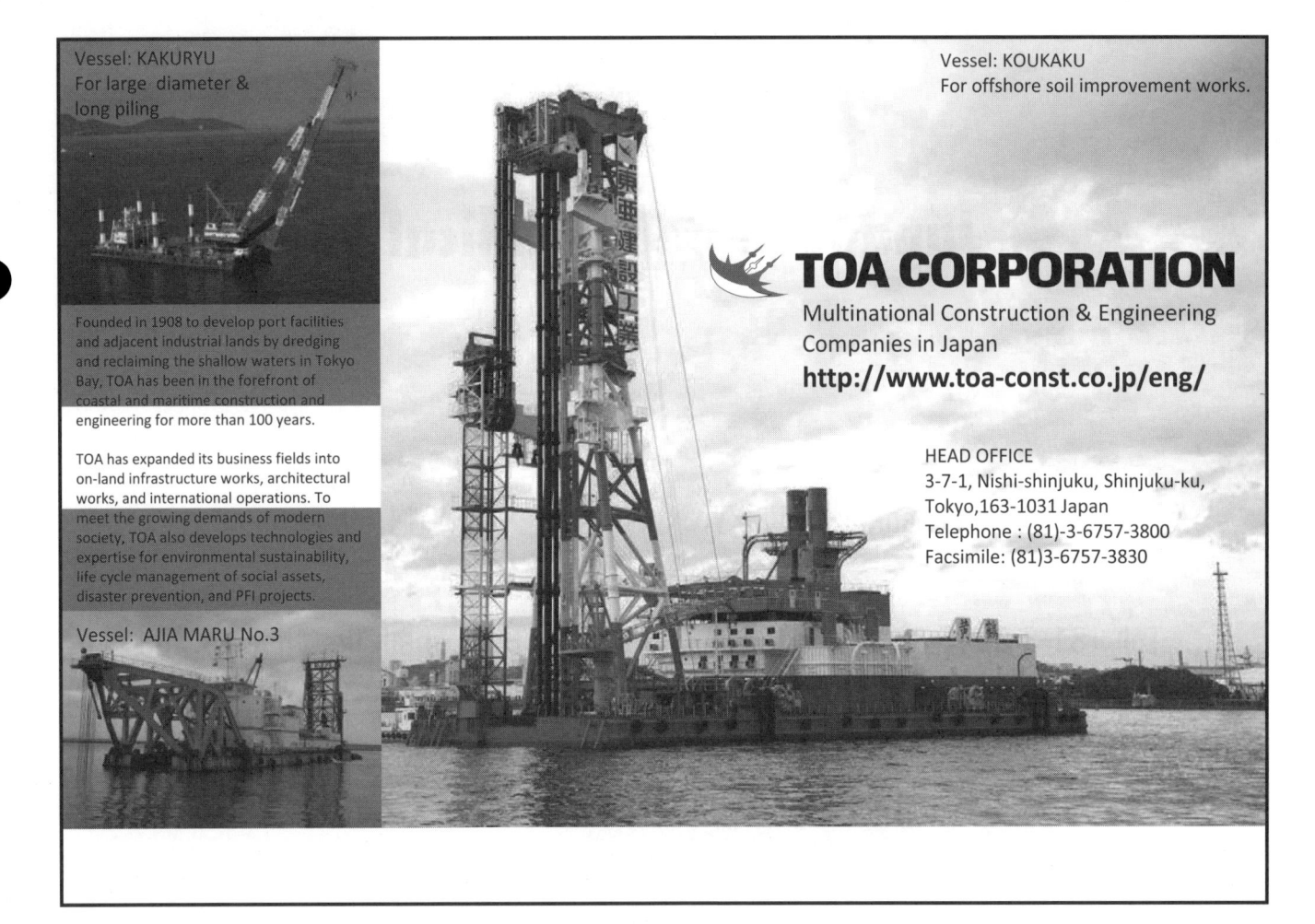

TAKENAKA

Takenaka Carpentry Tools Museum
design & construction: Takenaka Corporation

TAKENAKA CORPORATION http://www.takenaka.co.jp/takenaka_e/

Vessel: KAKURYU
For large diameter & long piling

Founded in 1908 to develop port facilities and adjacent industrial lands by dredging and reclaiming the shallow waters in Tokyo Bay, TOA has been in the forefront of coastal and maritime construction and engineering for more than 100 years.

TOA has expanded its business fields into on-land infrastructure works, architectural works, and international operations. To meet the growing demands of modern society, TOA also develops technologies and expertise for environmental sustainability, life cycle management of social assets, disaster prevention, and PFI projects.

Vessel: AJIA MARU No.3

Vessel: KOUKAKU
For offshore soil improvement works.

TOA CORPORATION

Multinational Construction & Engineering Companies in Japan
http://www.toa-const.co.jp/eng/

HEAD OFFICE
3-7-1, Nishi-shinjuku, Shinjuku-ku,
Tokyo,163-1031 Japan
Telephone : (81)-3-6757-3800
Facsimile: (81)3-6757-3830

Leaving to the future, society coexisting with nature

WESCO brings together its technical, creative and practical skills with integrated organizational strength to build a comfortable future for all of us.

Surveys
Surveys
Geographical information
Compensation surveys
Ground surveys
Environmental planning/surveys

Design
Civil engineering design
Agriculture and forestry promotion
Architectural design
Environmental design

Infrastructure management
Urban planning
PPP/PFI
Asset management
Orderer support system

For more details about our business, please visit our website.
http://www.wesco.co.jp/

General Construction Consultant
WESCO Inc.

Head Office : 2-5-35 Shimadahonmachi, Kita-ku, Okayama-shi, Okayama 700-0033 Japan
TEL : (+81) 86-254-2111 (Key) FAX : (+81) 86-253-2098

Japanese Geotechnical Society

4-38-2 Sengoku, Bunkyo-ku, Tokyo, 112-0011, Japan

Tel: +81-3-3946-8677   Fax: +81-3-3946-8678   E-mail: jgs@jiban.or.jp

Japanese Geotechnical Society Standards
Geotechnical and Geoenvironmental Investigation Methods Vol. 3

Published by
The Japanese Geotechnical Society
4-38-2 Sengoku, Bunkyo-ku, Tokyo 112-0011, Japan
E-mail: jgs@jiban.or.jp
URL: https://www.jiban.or.jp/e/

©2018 The Japanese Geotechnical Society
All rights reserved. This book, or parts thereof, may not be reproduced in any form or by any means electronic or mechanical, including photocopying, recording or any information storage and retrieval system now known or to be invented, without written permission from the publisher.

ISBN978-4-88644-825-5

Distributed by
Maruzen Publishing Co., Ltd.

Printed in Japan
by TD Planners Inc.

JAPANESE GEOTECHNICAL SOCIETY STANDARDS
—GEOTECHNICAL AND GEOENVIRONMENTAL INVESTIGATION METHODS—
Errata Sheet

2017/4/26

Vol.	Standards	Page Location	Incorrect	Correct
1	JIS A 1221:2013	p.4 6, c), line 3	$N_{sw} = 100/L\ N_s$	$N_{sw} = N_a/L$
1, 2	Review article	p.2 Table 1, Loading tests	JIS A1215 Vol.3	JIS A 1215 Vol.2
1, 2	Review article	p.2 Table 1, Loading tests	JIS A1222 Vol.3	JIS A 1222 Vol.2